T0350179

Contract Administration and Procurement in the Singapore Construction Industry

Contract Administration and Procurement in the Singapore Construction Industry

Lim Pin

B Sc (Building)(Hons), LLB(Hons)(London)
Advocate & Solicitor, Barrister
Associate Professor
Department of Building
School of Design & Environment
National University of Singapore

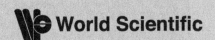

World Scientific

NEW JERSEY · LONDON · SINGAPORE · BEIJING · SHANGHAI · HONG KONG · TAIPEI · CHENNAI · TOKYO

Published by

World Scientific Publishing Co. Pte. Ltd.
5 Toh Tuck Link, Singapore 596224
USA office: 27 Warren Street, Suite 401-402, Hackensack, NJ 07601
UK office: 57 Shelton Street, Covent Garden, London WC2H 9HE

Library of Congress Cataloging-in-Publication Data
Names: Lim, Pin, author.
Title: Contract administration and procurement in the Singapore construction industry /
    Pin Lim (Nus, Singapore).
Description: New Jersey : World Scientific, 2016.
Identifiers: LCCN 2016028192| ISBN 9789813148031 (hardback : alk. paper) |
    ISBN 9789813148048 (pbk. : alk. paper)
Subjects: LCSH: Construction contracts--Singapore.
Classification: LCC KPP85.6.B84 L55 2016 | DDC 343.595707/8624--dc23
LC record available at https://lccn.loc.gov/2016028192

British Library Cataloguing-in-Publication Data
A catalogue record for this book is available from the British Library.

Desk Editor: Amanda Yun

Typeset by Stallion Press
Email: enquiries@stallionpress.com

Printed in Singapore

To Charlene (*mm*)

Preface

Contract Administration and Procurement is an important discipline in the management of cost, time and quality in construction projects. This book is written for the following:

Developers
Contractors
Project Managers
Architects
Engineers
Quantity Surveyors
Consultants and professionals in the construction industry
Students in project and facilities management, architecture, engineering and other allied courses

There is a dearth of books on this particular subject in the Singapore construction industry. Most students, including myself when studying for my degree in Building, do not have a textbook or materials providing instructions in contract administration. We picked up the skill whilst on the job. For example, how does one value the works on site for the purpose of making progress payments to the Contractor? As a young Quantity Surveyor, I accompanied my senior to site and was taught the rudiments of estimating the value of work done on site.

Although nothing beats learning through experience, this book endeavours to show how valuation of Payment Claims is made, how valuation of variations is arrived at, how the Final Account of a project is derived, and many other contract administration processes. Many examples and processes mentioned in this book are geared towards practice in the construction industry to benefit professionals and give students a foretaste prior to entering the industry. Consequently, draft letters, notices, certificates and other documents are provided in the book to help readers easily understand the processes. However, readers should not use such drafts as models or templates in their projects. Letters and other documents in respect of projects should be drafted in pursuit of a specific interest and in accordance with the facts and requirements in their specific circumstance(s). It should also be noted that there are many ways to draft letters and documents for a specific purpose. There is no one fixed way as the facts and circumstances may differ from project to project. And letters and documents should take into account the relevant facts and circumstances, as these are important factors in the pursuit of a specific interest. Similarly, the various flow-charts, time-lines and other diagrams are not "cast in stone" but allow for certain latitude, depending again on the facts and circumstances.

The contract administration processes take into account various provisions of the Singapore Institute of Architects (SIA) Measurement Contract 9th Ed, 2010, Public Sector Standard Conditions of Contract for Construction Works (PSSCOC) 7th Ed, 2014, and the Building and Construction Industry Security of Payment Act (SOP Act).

Disclaimer

Draft letters, notices, certificates and other documents in this book ("Drafts") serve as examples and illustrations for understanding only. They are not models or templates to be used in actual projects.

The author does not warrant that the Drafts are suitable for use in projects or any other circumstances.

Acknowledgments

I am grateful to Mr Winston Hauw, Partner, Rider Levett Bucknall, Singapore, for his advice and kindness, and Mr Wong Kin Hoong, Director, Kingsmen Exhibits Pte Ltd, for his invaluable assistance with materials.

I am also thankful to Associate Professor Daniel Wong Hwee Boon for his kindness in providing materials, proof-reading and correcting errors in a few draft chapters of this book. If there are any errors in the book, they are wholly mine and I take full responsibility for them.

I record my special thanks to Ms Amanda Yun, Senior Editor, World Scientific Publishing, Singapore for her help in the publication of the book and her team's effort in getting the book out in print in double quick time for students in August 2016. Last but not least, I thank my wife, Charlene, whose support is indispensable and a very present help indeed.

Lim Pin
B Sc (Building)(Hons), LLB(Hons)(London)
Advocate & Solicitor, Barrister
Associate Professor
Department of Building
School of Design & Environment
National University of Singapore
1 August 2016

About the Author

 Lim Pin is an Associate Professor in the Department of Building, School of Design and Environment, National University of Singapore. He is also the author of *Elements of Construction Law in Singapore.*

He graduated with a B.Sc. (Building)(Hons) from the National University of Singapore. He started his career as a Quantity Surveyor in the Lands and Estates Organization of the Ministry of Defence where he was fully engaged in contract administration work and procurement in construction projects. Subsequently, he practised as an advocate and solicitor in construction arbitration and commercial litigation. He was also legal counsel to corporations owning and managing properties such as Sentosa Development Corporation and also oversaw legal matters in property management in Suntec City.

Contents

CHAPTER 1

Introduction

For an introductory chapter, the Author wishes to provide a brief over-view of the book. More detailed analysis will be discussed in subse-quent chapters. This chapter will discuss the following:

1.1 Parties in a Construction Project
1.2 Roles of and Relationship Between Parties
1.3 Time-Line in a Construction Project
1.4 Pre-contract Administration
1.5 Post-contract Administration

1.1 Parties in a Construction Project

The main parties involved in a construction project are as follows:

1.1.1 Owner or Developer
1.1.2 Project Manager
1.1.3 Architect, Civil & Structural Engineer ("C&S Engineer"), Quantity Surveyor ("QS"), Mechanical & Electrical Engineers ("M&E Engineers") and other Consultants
1.1.4 Main Contractor
1.1.5 Sub-Contractors
1.1.6 Suppliers

1.1.1 Owner or Developer

An owner of land ("Owner") has the power to decide whether to build on his land. It may be for the purpose of building a residential house or developing a block of apartments for sale or for other commercial uses. The Owner may be a private sector owner or public sector owner of properties as follows:

1.1.1.2 Private sector

Person(s) or companies may own properties. They may decide to build a house on the land. A developer of condominiums is also a private owner. They may build condominiums for sale with a commercial objective. Examples of developers in Singapore are Far East Organization, CapitaLand Limited, and City Development Limited.

1.1.1.3 Public sector

The Government is also an owner of properties. The State may own properties directly or through many of the Government-supervised statutory boards. Examples of property-owning statutory boards are the Housing Development Board ("HDB"), Land Transport Authority ("LTA"), and Sentosa Development Corporation.

1.1.2 Project Manager

Traditionally, the Architect, being the leader of the consultancy team, doubles up as the *Project Manager* in the management of the project. Due to present day's large and complicated projects, there is usually a Project Manager. He is accountable to the Owner in managing the whole project to ensure that it is completed on time and within budget. The Project Manager may be qualified as an Architect, an Engineer, a Quantity Surveyor or other professions as long as the person is competent and experienced in the management of construction projects. The Architect and other consultants take direction from the Project Manager.

There may be Project Managers in the Private sector or Public sector as follows:

1.1.2.1 Private sector

There are Project Managers employed in-house in developer companies e.g. Far East Organization, CapitaLand Limited. They act in the interest of their developer employers to ensure that the external consultants who were out-sourced and contractors who were engaged carry out their duties properly.

Project Managers may also provide their services to developers in their capacity as out-sourced firms, e.g. CPG Corporation. These firms provide project management services to developers in consideration for a fee.

1.1.2.2 Public sector

There are Project Managers employed in-house in statutory boards, e.g. HDB, LTA, to manage and oversee projects. The statutory boards may also out-source the project management services to other project management firms.

1.1.3 Architect, Engineers, Quantity Surveyor and Other Consultants

Professionals, such as Architects, Engineers and Quantity Surveyors, known as *Consultants* provide services in design, supervision of works, cost management and other professional services. Architects, Engineers and Quantity Surveyors may be employed in the private sector in developer companies or practice respectively in architectural, engineering, quantity surveying or multi-disciplinary firms. A multi-disciplinary firm is a firm which provides different professional services such as architectural, engineering, quantity surveying and other consultancy services within the same firm.

Such professionals may also be employed in the public sector, e.g. HDB.

1.1.3.1 Architect

Usually, the Owner appoints an Architect first to provide the concept design for the Owner's approval. Traditionally, the Architect is the leader of the consultancy team consisting of the Architect (as team leader), the Civil and Structural Engineer, the Mechanical and Electrical Engineers, and the Quantity Surveyor.

1.1.3.2 Civil and structural engineer

Upon approval of the concept design, the Architect would forward the design to the Civil and Structural Engineer ("C&S Engineer"). The C&S Engineer has to make sure that the design is structurally sound by designing a system of piles, reinforced concrete beams and columns to carry the load of the building itself and fixtures, known as the *dead load*, and the movable load such as furniture and persons in the building, known as the *live load*.

1.1.3.3 Quantity surveyor

Upon approval of the concept drawings by the Owner (mentioned in para. 1.1.3.1), the Quantity Surveyor ("QS") would be requested to provide an estimate for the cost of the construction project. Naturally, the estimate should meet with the Owner's approval before proceeding with the detailed design.

The estimate approved by the Owner will usually be the budget set by the Owner for the construction and completion of the project. It would be the responsibility of the consultancy team to ensure that the cost of the project is kept within the budget set by the Owner.

1.1.3.4 Mechanical and electrical engineers

Upon the Owner's approval of the concept drawings and cost estimate, the Mechanical and Electrical Engineers ("M&E Engineers") would also be appointed.

The M&E Engineers would be required in the design and supervision of air-conditioning, lighting, pumps and other mechanical and electrical installations.

1.1.4 Main Contractor

The consultancy team has to draft and prepare the specifications, drawings and other tender documents to engage a *Contractor*. The Contractor engaged to construct and complete the whole construction project is called the "Main Contractor". Although the Main Contractor may be responsible for delivering the whole of the completed works to the Owner, he may not carry out some of the specialist works e.g. mechanical and electrical works.

1.1.5 Sub-Contractors

The Main Contractor would sub-contract some of the specialist works to other contractors (called "Sub-Contractors").

Traditionally, the air-conditioning works and electrical works are sub-contracted to Air-conditioning and Electrical Sub-Contractors respectively as these are specialist works. Although such specialist works are sub-contracted to Sub-Contractors, the Main Contractor remains responsible and liable to the Owner for the satisfactory completion of these specialist works.

1.1.6 Suppliers

The Main Contractor and Sub-Contractors purchase building materials and equipment from *Suppliers* and then install such materials and equipment into the construction work. Sometimes, Suppliers may provide value-added services by installing the materials into the works, e.g. supplying and placing the concrete by a ready-mixed concrete supplier.

1.2 Roles of and Relationships Between Parties

The following figure shows an example of the roles of, and relationships between different parties in a construction project.

Please refer to Fig. 1.1 in respect of the roles and relationship described below.

1.2.1 Owner/Employer

Through contractual relationships, the Owner (also known as the *Employer*) engages the following:

- Architect;
- C&S Engineer;
- QS;
- M&E Engineers;
- Main Contractor.

The contractual relationships are shown in Fig. 1.1 by the arrow, ———▶.

However, the Owner (e.g. a developer) may employ other Architects, Engineers, QS and other professionals within the Owner's organization. On behalf of the Owner, these professionals monitor and give instructions to Consultants in respect of projects to ensure that the project is progressing in good order.

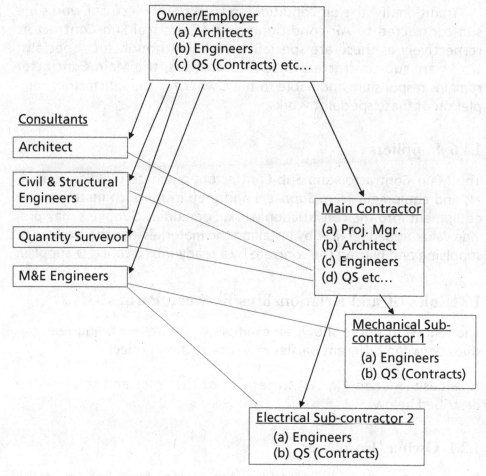

Fig. 1.1   Roles of and relationship between Parties.

In addition to the duties described, professionals employed by the Owner also undertake projects where Consultants are not required, e.g. minor repair and maintenance work.

1.2.2 Consultants

Generally, the duties of Consultants to the Owner are to:

- design the project,
- supervise the works,
- exercise cost control,

- ensure good quality in workmanship of the Contractors,
- ensure completion on time by Contractors.

Hence, Consultants have to monitor, direct and instruct the works by the Main Contractor and Sub-Contractors as the work progresses. This is shown in Fig. 1.1 by the straight line _____.

Note that there are no contractual arrow lines but only straight lines (supervisory function) between Consultants and Contractors in Fig. 1.1.

The duties of the Architect may include the following:

- Planning & designing the project;
- Planning, monitoring and ensuring that the Contractors keep to the time-schedule;
- Ensuring good quality in workmanship of the Contractor;
- Supervising the works according to specification, drawings and statutory requirements.

As explained, there is no contractual relationship (no arrow lines) between the Architect and the Contractors. The Architect has a supervisory function over the Main Contractor (denoted by the straight line in Fig. 1.1). Despite the lack of contractual relationship between the Architect and Main Contractor, the Main Contractor is obliged to take instructions from the Architect due to the contractual obligations between the Owner and Main Contractor that require the Main Contractor to comply with the Architect's instructions.

The primary duty of the C&S Engineer is to ensure that the structure is designed and completed in accordance with the requirement of the relevant statutes, codes of practice and standards. As seen in the straight line in Fig. 1.1, the C&S Engineer supervises the Main Contractor's works in the construction and completion of the structure.

The QS's duties are to compile and collate:

(a) the specifications and drawings,
(b) tender instructions,
(c) the form of tender and other tender documents, for the purpose of calling for tender by Contractors.

Tenderers would be pre-qualified based on previous performance, track record of previous projects and financial value of projects previously undertaken. The Building & Construction Authority provides a register of contractors[1] which may be used for the purpose of inviting Contractors to tender.

Thus, the QS administers the tendering process as follows:

(i) recommends appropriate tenderers to tender for the project,
(ii) sends letters of invitation to tender,
(iii) fixes the date for site show-round during which the Architect (and other consultants) will brief prospective tenderers of the works at the site,
(iv) arranges with tenderers for collection of tender documents upon payment of tender deposits,
(v) evaluates and recommends (together with the Architect) a successful tenderer to the Owner for acceptance upon submission of tenders.

As seen by the straight line in Fig. 1.1, the QS administers the construction contract by interacting with the Main Contractor in the valuation of progress payments, valuation of variation, and finalization of accounts at the completion of the construction works.

The duties of the M&E Engineers are to design and supervise the mechanical and electrical works. Naturally, such works include installation of lighting, air-conditioning and mechanical ventilation. As seen by the straight line in Fig. 1.1, the M&E Engineers would usually supervise the Electrical and Mechanical Sub-contractors' works.

1.2.3 Main Contractor and Sub-contractors

The roles of the professionals employed by the Main Contractor and Sub-contractors are as follows:

- Perform the role of project management for the Main Contractor or Sub-contractor (in most cases);
- Ensure that the works are completed in time, in good quality, within budget and in accordance with the construction contract;
- Provide minor design and preparation of shop drawings.

---

[1] http://www.bca.gov.sg/BCADirectory/Classification/Details/128.

As seen in Fig. 1.1, these professionals (usually a Project Manager or Engineer) liaise with the Consultants and ensure that the Consultants' instructions are carried out. They plan and co-ordinate the works such that they are completed in a timely manner according to the contract. Hence, their role is an important one in the construction process.

Also as seen in Fig. 1.1, a Sub-Contractor has no contractual relationship with the Owner (no arrow line between Owner and Mechanical Sub-Contractor 1, no arrow line between Owner and Electrical Sub-Contractor 2). As such, Sub-Contractors are not liable to the Owner for the works. In other words, if a Sub-Contractor constructs the sub-contract works defectively, such a Sub-Contractor will not be liable for breach of contract to the Owner for the defective works. The Main Contractor will be liable to the Owner for the defective works. The contract between the Owner and Main Contractor requires the Main Contractor to be liable and responsible for the Sub-Contractors' defective works.

Similarly, by virtue of the sub-contract between the Main Contractor and Sub-Contractor, the Sub-Contractor will be liable for breach of contract to the Main Contractor for the Sub-Contractors' defective works.

Sub-Contractors may be classified as follows:

- Designated Sub-Contractors
  A *Designated Sub-Contractor*[2] is one who has been expressly identified in the contract between the Main Contractor and the Owner ("Main Contract"). The Owner has, in the Main Contract, required the Main Contractor to engage the specific Sub-Contractor to perform certain specified works.

- Nominated Sub-Contractors
  A *Nominated Sub-Contractor*[3] is one who has not been identified in the Main Contract. But the Main Contract has provided that part of the Main Contract works would be performed by a Sub-Contractor to be nominated by the Owner at a later date.

---

[2] SIA Measurement Contract 9th Edition, cl 28(1)(a).
[3] SIA Measurement Contract 9th Edition, cl 28(1)(b).

- Domestic Sub-Contractors
  A *Domestic Sub-Contractor* is one who is engaged by the Main Contractor to perform part of the Main Contract's works. Usually, the Owner has no part in the selection of the Domestic Sub-Contractors.

1.3 Time-line in a Construction Project

A simplified time-line of a construction project may be represented in Fig. 1.2.

The two main stages of Pre-Contract and Post-Contract on the time-line are set apart by the *Acceptance of Successful Tender*. Generally, activities prior to the Acceptance of Tender are known as *Pre-Contract Administration*, and activities after the Acceptance of Tender are known as *Post-Contract Administration*.

1.4 Pre-Contract Administration

Some of the main activities for Pre-Contract Administration may be referred to in Fig. 1.2.

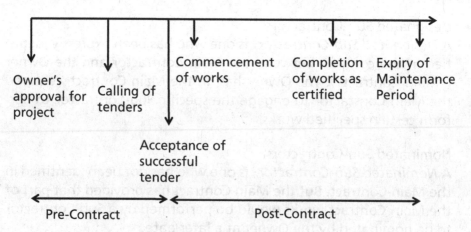

Fig. 1.2   Time-line in a construction project.

1.4.1 Owner's Approval for Project

The Owner's approval is naturally required in any project. The Owner has to approve:

- the concept design for the overall appearance of the project,
- the budget for the project,
- time for completion.

Upon the Owner's approval, the Architect would develop more detailed plans for submission to the relevant Authority for approval (e.g. URA). Upon the Authority's approval, the tender drawings and documents would be prepared and finalized for tendering.

1.4.2 Calling Tenders

The traditional method of choosing a contractor to carry out the works is by way of a competitive tender (or bid or offer). The different tendering systems will be discussed in a later chapter. Usually, the successful tenderer (tenderer who is awarded the construction project) is one who is the lowest tenderer and complies with all the requirements of the tender.

1.4.3 Acceptance of Successful Tender

In a construction tender, a tenderer (usually a contractor) offers a price which the tenderer wants to be paid in order to carry out and complete the construction works. If the Owner accepts the price offered by posting a *Letter of Acceptance* (or *Letter of Award*), a contract comes into effect between the Owner and the successful tenderer.

1.5 Post-Contract Administration

Once a Letter of Acceptance is posted, the time-line enters the Post-Contract Administration stage. Some of the main activities in respect of Post-Contract Administration may be referred to in Fig. 1.3.

1.5.1 Issue of Letter of Acceptance to Completion of Works

The Contractor may commence work on the *Contract Commencement Date* stated in the contract or otherwise stipulated. Only when

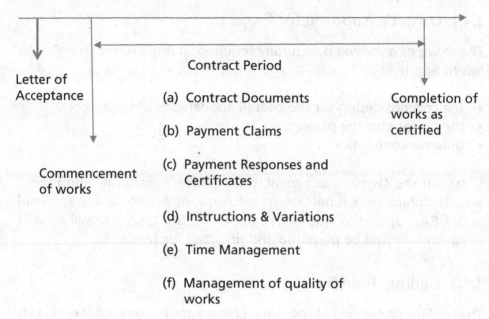

Fig. 1.3   Time-line from issue of letter of acceptance to completion of works.

the Architect (or, in the case of PSSCOC form, the Superintending Officer) has *certified* the works as completed is the work deemed to be completed in accordance with the contract. As may be seen in Fig. 1.3, there are many Post-Contract Administration activities, some of which will be discussed in greater detail in later chapters.

A brief description of Post-Contract Administration activities is provided as follows:

1.5.1.1 Contract documents

Upon acceptance of the tender, the *Contract Documents*[4] would be prepared and signed by the parties. The documents generally consist of the following:

(a) The Articles of Contract;
(b) The Conditions of Contract;
(c) Drawings;
(d) Specifications;

---

[4] SIA Measurement Contract 9th Edition, Art 6.

(e) Bills of Quantities (if any);
(f) Such other letters, documents such as Form of Tender, Letter of Invitation etc., as parties may agree to and attach as part of the Contract Documents.

Prior to being part of the Contract Documents, many of the documents were part of the Tender Documents available to the Contractor for preparing the tender.

1.5.1.2 Payment claims

By virtue of the Building and Construction Industry Security of Payment Act ("SOP Act"), Contractors submit *Payment Claims*[5] to the Owners, usually monthly. The Payment Claim is the Contractor's claim for his work based on work done and material delivered to site.

1.5.1.3 Payment responses and certificates

In response to each of the Contractor's Payment Claims, the Owner will provide a *Payment Response*.[6] The Payment Response is the Owner's assessment (through the advice of the QS) of the amount of work done, value of material on site and value payable.

Arising from the contract, the Architect would issue an *Interim Certificate* for payment (also known as "Payment Certificate" or "Progress Payment Certificate").[7]

1.5.1.4 Instructions and variations

During the progress of the works, the Architect may give instructions to the Contractor to carry out the works. The Contractor may be entitled to additional payment in respect of some of these instructions. In other instances where an Architect gives a *Direction*, no additional payments are allowed.[8]

---

[5] SOP Act, s 10. Refer to Chapter 5.
[6] SOP Act, s 11. Refer to Chapter 6.
[7] SIA Measurement Contract 9th Edition, cl 31(3). PSSCOC 7th Edition, cl 32.2(1).
[8] SIA Measurement Contract 9th Edition, cl 1(2).

Generally, those instructions which require changes to the original contract intention are known as *Variations* to the contract. For example, the Architect issues an instruction requiring the change of floor tiles (in the contract) to carpet. This is known as a *Variation Order* and is given in the form of an *Instruction*[9] in accordance with the contract.

1.5.1.5 Time management

Consistent with prudent time management, the contract should provide for time for completion of the whole of the works. However, completion by the date stipulated in the contract is rare due to many unexpected events (or even expected events) during the construction period.

Hence, most construction contracts do not require absolute fulfilment of completion by a stipulated completion date. Instead, non-completion by a stipulated date in the contract is classified as under an *Extension of time* ("EOT") or *Delay*. If the non-completion was due to events expressly provided in the contract,[10] EOT will be granted. If non-completion was due to events not extendable by EOT in the contract, then, the Contractor is said to be in Delay. The Owner may impose *Liquidated Damages* ("LD") arising from Delays. If the Contractor is granted EOT, no LD may be imposed for the duration of the EOT.

Common standard forms provide, among other clauses, that Contractors may be granted EOT for adverse weather or for additional work instructed by the Architect. These events would delay the stipulated completion date in the contract but they are through no fault of the Contractor. Therefore, EOT will be granted with no imposition of LD.

On the other hand, if the Contractor could not complete the work due to poor co-ordination and planning of construction work, the Contractor will be said to be in Delay. Poor co-ordination and planning leading to a delay in the works are due to the fault of the

---

[9] SIA Measurement Contract 9th Edition, cl 1. PSSCOC 7th Edition, cl 19.1, 19.2. Refer to Chapter 7.
[10] SIA Measurement Contract 9th Edition, cl 23.1. PSSCOC 7th Edition, cl 14.2.

Contractor. The contract does not in such an instance provide for EOT. Hence, the Contractor would be in delay and LD may be imposed by the Owner.

1.5.1.6 Management of quality of works

Defects are not uncommon in construction works. These defects may occur during the period of construction or when the whole of the works are completed and handed over to the Owner. Some defects are caused by Contractors and others are not due to the fault of Contractors.

If defects were caused by Contractors or within the responsibility of Contractors to repair, Architects may give a Direction[11] or Instruction[12] for repair (without cost to the Owner). Further, when the works are wholly completed and handed over to the Owner, there are further requirements for repair during the Maintenance Period or Defects Liability Period.

1.5.2 Completion of Works to Issue of Final Certificate

Subsequent to the Completion of works as certified by the Architect, the Contractor is responsible to repair defects, complete any outstanding works during the *Maintenance Period* and *Finalization of Account* as shown in Fig. 1.4.

1.5.2.1 Maintenance period

The contract usually provides for the Maintenance Period[13] (see Fig. 1.4) or *Defects Liability Period*[14] to commence on the date of Completion of works as certified by the Architect. During the Maintenance Period (or Defects Liability Period), the Contractor shall be required to repair defects without cost to the Owner. Usually, the contract provides for a 12-month Maintenance Period.

---

[11] SIA Measurement Contract 9th Edition, cl 11(3).
[12] PSSCOC 7th Edition, cl 10.7.
[13] SIA Measurement Contract 9th Edition, cl 27(1).
[14] PSSCOC 7th Edition, cl 1.1(k).

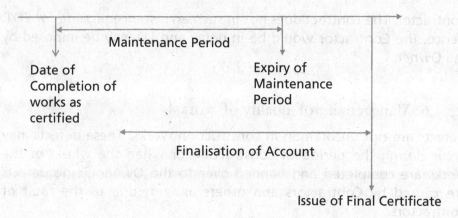

Fig. 1.4   Time-line for the maintenance period and finalization of account.

1.5.2.2 Finalization of account

The Finalization of Account document is a document providing for the valuation of all addition and omission of works and other adjustments to arrive at an adjusted *Contract Sum*. Then, the *Final Payment* would be made to the Contractor based on the adjusted Contract Sum.

The submission of Contractor's *Final Account* takes place during the Maintenance Period (SIA Contract) or Defects Liability Period (PSSCOC Contract). Upon receipt, the Owner's QS would determine his own assessment of the Final Account.

The *Final Certificate*[15] (or Final Account Certificate) would certify the *Final Payment* to the Contractor.

---

[15] SIA Measurement Contract 9th Edition, cl 31(12)(c). PSSCOC 7th Edition, cl 32.5(7).

CHAPTER 2

Procurement

This chapter will discuss the following:

2.1 Meaning of Procurement
2.2 Key Concepts to Good Procurement
2.3 Methods of Sourcing a Contractor

2.1 Meaning of Procurement

Procurement refers to the process of purchasing goods or services. A common method of purchasing construction service is through calling of tenders.

2.2 Key Concepts to Good Procurement

Singapore is party to the World Trade Organization's Agreement on Government Procurement. As such, the Singapore Government's procurement framework is aligned with international standards and obligations. The following are the key concepts in procurement and the tendering process that are also to some degree observed in the private sector:

(a) Transparency

Transparency is desired so that the tender process may be seen to be fair and reasonable. Transparency also minimizes allegation of biasness in favor of certain vendors or contractors. Thus, the terms to be eligible to tender are made known to all interested parties. In this way, transparency is maintained.

(b) Open and fair competition

In order to get a competitive tender (i.e. lowest price for purchasing goods or services), there must be open and fair competition.

The tender should be open to as many eligible tenderers as possible to avoid allegations of discrimination and to encourage competition. The tenderers would be competing on the same terms and conditions in as fair a manner as possible. Thus, tenderers are fully aware that there is a *level playing field,* where every tenderer is to tender on the same terms. They are also aware that the successful tenderer would be the one to submit the most competitive tender. In this manner, the tender system motivates tenderers to submit a competitive tender for good value.

(c) Value for money

Value for money means optimizing between benefits and costs in the purchase of products or services. Sometimes the balance between benefits and costs may not result in award of the tender to the lowest tenderer. This encourages tenderers to tender not only competitively but, where required, innovatively, proposing better quality products or services at marginal increase in cost.

2.2.1 Government Procurement Act ("GPA")

2.2.1.1 What is GPA?

The GPA[1] is an Act enacted to give effect to Singapore's international treaty obligations under the Agreement on Government Procurement 1994 established under the World Trade Organization.

2.2.1.2 Why GPA?

The Agreement on Government Procurement treaty with other countries allows the Government of Singapore and other governments of participating countries to enter into a more 'level playing field' in Government Procurement. This allows our local business entities to supply goods and services to the governments in these participating countries and therefore open these overseas markets to our suppliers.

---

[1] Cap 120.

The corollary is that Singapore likewise must open our Government procurement to these overseas participating countries. Overseas suppliers have similar opportunities of supplying goods and services to the Singapore Government.

The overall scheme of more open cross border supply of goods and services in government procurement is achieved through the treaty and the Act.

2.2.1.3 Who are involved in GPA?

The participating countries are provided in the Government Procurement (Application) Order.[2] Some of these are the countries in the European Union, USA and Canada.

2.2.1.4 Principles governing GPA

The Government Procurement Regulations 2014, a subsidiary legislation of the GPA provides as follows:

4.—(1) A reference in these Regulations to the principles of national treatment and non-discrimination is a reference to the following principles:

(a) *that the goods and services of a relevant State or a relevant Protocol State are not to be treated less favorably than Singapore goods and services or the goods and services of any other relevant State or relevant Protocol State;*

The above stipulates that the goods and services of relevant States and Protocol Relevant States are not to be treated less favorably than those in Singapore. Hence, this leads to the following principles in GPA:

- Openness and fairness — suppliers are given equal access to opportunities to compete so as to ensure that the best offers are received;

---

[2] G.N.No.S217/2002 Revised Edition 2004.

- Transparency — different stages of the procurement process should be available for viewing so as to ensure that there is fairness in the process of procurement and integrity in the system;
- Value for money — the tender need not be awarded to the lowest tenderer; the tenderer who offers optimal value in the balance between benefit and cost should be favored.

2.2.1.5 Singapore's compliance with GPA

The Singapore Government system for procurement is through GeBIZ, an electronic web portal. Participating countries' suppliers may register with GeBIZ. Therefore, such suppliers have access to our Government tenders and are able to compete with our local suppliers on equal basis.

2.3 Methods of Sourcing a Contractor

An important part of the procurement process is the sourcing of a suitable Contractor. There are many methods of sourcing a Contractor.

Traditionally, the Owner (or Employer) engages a team of Consultants (including the Architect). The Architect provides a design for the project. The design is implemented by the carrying out and completion of construction works. Typically, a procurement process involves the following stages, shown in Fig. 2.1 below.

2.3.1 Sourcing Contractors

Generally, the common method for sourcing Contractors is by way of *calling tenders*. The Owner's Consultants would prepare tender documents such as specifications and drawings for tenderers to bid competitively. Usually, the successful tenderer is the lowest bidder who

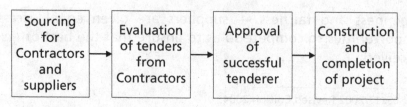

Fig. 2.1 Simple process chart for procurement of construction.

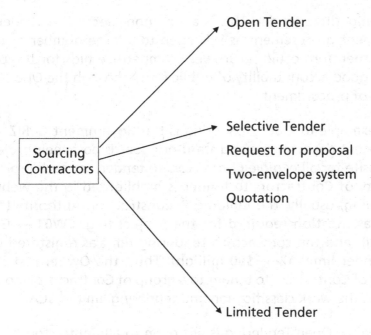

Fig. 2.2   Different methods of Tendering.

complies with the tender documents. Fig. 2.2 is a chart showing the different methods of tendering by Contractors.

There are three basic methods of tendering — Open, Selective, and Limited Tendering. A tender is an *offer* for which an acceptance by the Owner would effect a binding contract between the Owner and the successful tenderer. Hence, it is an effective method of sourcing Contractors by requiring them to tender (offer) a price to construct and complete the project. The Owner accepts the most suitable tender.

Through this method, the Owner has the advantage of evaluating and deciding on a suitable Contractor to do the project out of many tenderers.

2.3.1.1 Open tender

An *Open Tender* is a tendering exercise that is open for all to tender. The objective of an Open Tender is to have fairness in allowing all Contractors to tender. The successful tenderer is usually the most competitive tender. This is a common method of tendering in Government procurement as it is open to a large number of tenderers and therefore could secure very competitive bids for the project. There is good accountability of public funds through the Open Tender method of procurement.

An example of Open Tendering is the Government GeBIZ[3] where Government departments and statutory boards post their projects on the website for all *eligible Contractors* to tender. For each project, the eligibility of Contractors to tender is highlighted in the website by mentioning, usually, the Building & Construction Authority ("BCA") work classification required for the project (e.g. CW01 — General Building), and the contractor's tendering limit as registered in BCA (e.g. tender limit A2 — $90 million). Thus, the Owner restricts the number of Contractors to tender to a group of Contractors who qualify based on the work classification and tendering limit in BCA.

In reality, Open Tendering is not open to all Contractors to tender but only those that satisfy the criteria set by the Owner. This is necessary as Contractors who are not qualified or financially incapable of undertaking a particular type of project should not be tendering for such projects. For example, Contractors who specialize in minor renovation works ought not to tender for multi-million dollar projects and vice-versa.

2.3.1.2 Selective tender

In *Selective Tender*, a number of Contractors are invited to tender. The number invited may range from as few as 3 to exceeding 10 contractors. The objective is to invite Contractors whom the Owner and his Consultants are of the view are capable of carrying out and completing the job.

The difference between Open Tender and Selective Tender is that the tender in the former is open to all eligible Contractors to participate in tendering. In Selective Tender, however, the Owners and Consultants select a number of Contractors to be invited to tender.

---

[3] https://www.gebiz.gov.sg/

Hence, in Selective Tender, not all eligible Contractors may tender as only those selected and invited may do so.

In a Design-Bid-Build ("DBB")[4] project delivery system, where the Consultants provide the design and the Contractors bid a price to build the project, the selected tenderers are required to bid a price for the construction and completion of the project.

However, in the Design and Build ("D&B")[5] project delivery system, the selected tenderers are required to provide a design as well as a price for construction according to their design if awarded.

Hence, in Selective Tender, tenderers may be selected and invited to either:

(a) bid a price for construction only based on design provided by Consultants; or
(b) bid a price on the basis of providing design and construction.

In situations where Contractors are selected to provide design and construction, two methods may take place: Request for Proposal or Two-envelope System.

2.3.1.2.1 Request for Proposal

The *Request for Proposal* ("RFP") is a request by the Owner to selected tenderers to provide proposals based on certain broad criteria. For example, the process for procuring the Integrated Resort ("IR") developer at Sentosa was through RFP.[6] At the close of the RFP, there were three high quality proposals, each from a well-known developer. Each developer provided a design for the IR and its financial investment based on the criteria set by the Government.

After considering and evaluating the three proposals, the Government announced the award to the winner for the RFP.

---

[4] Refer to chapter 3.
[5] Refer to chapter 3.
[6] https://www.mlaw.gov.sg/news/speeches/remarks-by-dpm-prof-s-jayakumar-at-press-conference-on-award-of-integrated-resort-at-sentosa-8-dec.html

In the example above, the RFP winner was more than a Contractor. They were the developer and lessee for a fixed term of years in operating and managing the IR. The IR developer called separate tenders for construction of the IR in accordance with the winning design and proposal.

Thus, similar RFPs may be applied to construction tenders. Contractors may be selected and invited on the basis of providing proposals consisting of design, construction and financial requirements based on criteria set by the Owner. The Owner may choose the winning proposal. The successful Contractor would have to deliver the completed design and construction according to the proposal to the Owner at the end of the day.

2.3.1.2.2 Two-envelope system

The *Two-envelope System* is another selective tender method. It requires each tenderer to provide their tender in two envelopes. One envelope provides the technical proposal and the other, the finance proposal.

Based on rules of this method, all envelopes containing the technical proposal would be opened first. The technical proposals of the different tenderers would be evaluated and those tenderers whose proposals do not meet the expressed technical requirements would not be considered. Next, only the financial proposals of tenderers who met the technical requirements would be opened and evaluated. Naturally, the tenderer with the best financial proposal would also be one who qualified in his technical proposal.

Thus, the Two-envelope System may be applied in the design and build method of delivery. Each tenderer may be required to provide two envelopes: one for the proposed design and the other for the price of design and construction. The Owner may decide on the award after evaluating the two envelopes.

The rule of opening the technical proposal first followed by the financial proposal has its object of guiding the decision maker into a decision which would satisfy both the technical and financial requirements. The Author understands that such a rule is not strictly complied with and sometimes all envelopes are opened at the same time for evaluation.

A non-construction example in applying the Two-envelope System was in the tender for the development of the former NCO Club tender at Beach Road.[7]

Both the RFP and Two-envelope System are not common methods of procurement in Singapore although they may be considered based on the circumstances.

2.3.1.2.3 Quotation

Some works are too small in value to justify time, effort and resources to call a tender, e.g. adding two cubicles in an office. The cost of a few partitions, tables, chairs and office equipment may be worth a few thousand dollars but it may not be worth the while to draft specifications, provide drawings, and invest the time and effort in a tendering process.

Thus, quotations may be obtained from a few vendors as explained in para 2.3.4 Financial Limits in Corporate Procurement. Thereafter, the most suitable quotation would be accepted.

2.3.1.3 Limited tender

In *Limited Tender*, usually only one contractor is chosen and invited to tender for the project. This method of tendering is not commonly practiced as competition in pricing for the project is absent and the Owner may have to resort to negotiation with a single tenderer to obtain a lower price.

It is used where the product or services offered by the tenderer is specialized in nature and other contractors may not be able to offer the same. Examples are constructing a special ride for an amusement park and commissioning an artist to undertake and complete an artwork. For the sake of confidentiality, sometimes only one contractor is approached to carry out and complete certain military and defense installations.

---

[7] http://propertyhighlights.blogspot.sg/2007/10/singapore-govt-used-two-envelope-system.html

2.3.2 Pre-Tender, Tender and Post-Tender Process

Subject to variations, the following flow chart in Fig. 2.3 shows a typical flow process from the conception of needs to tendering and ending with completion of construction and maintenance of a project.

Needs	There are Public Sector needs such as:

There are Public Sector needs such as:

- Social needs, e.g. schools, community centers,
- Public amenities, e.g. parks,
- Infrastructure and Transportation, e.g. drainage, roads, Mass Rapid Transit (MRT), airport,
- Public housing, e.g. Housing Development Board (HDB).

There are Private Sector (business) needs such as:

- Manufacturing industry, e.g. factories,
- Retail,
- Hotel,
- Entertainment,
- Private housing.

Appointment of Consultants

Upon Owner's decision to build to satisfy the needs, Consultants in the Public Sector may be appointed through:

- Design competition;
- Employment of in-house professionals within the public sector organization to design and supervise the works.

In the Private Sector, Consultants may be appointed through:

- Design competition,
- Direct appointment,
- Employment of in-house professionals within the Owner's organization.

The following Consultants are usually appointed through a contract:

- Architect,
- Quantity Surveyor ("QS"),
- Civil and Structural Engineer ("C&S Engineer"),
- Mechanical & Electrical Engineers ("M&E Engineers").

Fig. 2.3   Flow-chart for Pre-Tender, Tender and Post-Tender process.

Fig. 2.3 (*Continued*)

The fee of Consultants is usually based on a percentage of the cost of construction of the whole project.

Concept design and preliminary estimate	Upon appointment, the Architect will provide sketches showing the concept design of the building for the Owner's approval. The QS will provide an estimate of the construction cost for approval by the Owner.

Upon appointment, the Architect will provide sketches showing the concept design of the building for the Owner's approval. The QS will provide an estimate of the construction cost for approval by the Owner.

Owner's approval of concept design and preliminary estimate

The Owner will deliberate, request changes and approve the concept design and preliminary estimate of cost. This is an important decision as it gives the Consultants the mandate to start the procurement process through calling of tender in most cases.

Preparation of drawings, specifications and tender documents by the Consultants

Upon the Owner's approval of the concept design and preliminary estimate of cost, the following will likely take place:

- Architect will develop the concept design;
- C&S Engineer will design and provide the structural drawings;
- M&E Engineers will design and provide the drawing for the mechanical and electrical installations;
- QS will collate and compile the Tender Documents.

The objective of the above is to draft the Tender Documents for contractors to tender for the works. The Tender Documents will usually consist of the following documents:

- Form of Tender;
- Conditions of Contract;
- Drawings;
- Preliminaries;
- General specifications;
- Particular specifications;
- Breakdown of cost;
- Summary of Tender;
- Other requirements and information to the contractor for purpose of tendering.

(*Continued*)

Fig. 2.3 *(Continued)*

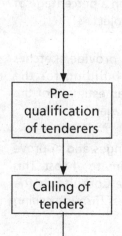

When the Tender Documents are ready, tenders will be called. As described earlier, there are different ways in calling of tenders:

- Open Tender — all eligible tenderers are welcomed to tender for the project;
- Selective Tender — a few tenderers would be invited to tender for the project;
- Limited Tender — negotiate the price and other terms with usually one supplier or contractor e.g. confidential tenders for defense and security installations;
- Government Procurement — tenders are called through GeBIZ in respect of Government departments and statutory boards.

A common method for the calling of tender is by way of Selective Tender. Frequently, the Consultants on behalf of the Owner will publish a notice through the media for interested contractors to forward:

- their track record of past projects,
- their financial value of past projects,
- their company particulars, e.g. paid-up capital, company particulars according to business printout by ACRA, and
- other information required

for shortlisting to be invited to tender for the project. This process is called *pre-qualification of tenderers*.

From the above information provided, the Consultants would shortlist and invite suitable tenderers to tender for the project. The number of tenderers invited should be such as would generate a healthy competition so as to return a competitively low price for the Owner. Hence, it makes little sense to invite only two or three tenderers to tender as such a small number of tenderers may not be competitive in the pricing.

Government departments and statutory boards must call tenders (and quotations) through GeBIZ if they exceed a certain financial value for the purchase of goods or services. GeBIZ will be discussed in a later section.

(Continued)

Fig. 2.3   (*Continued*)

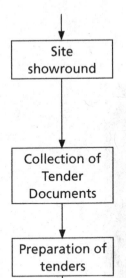

Site showround

In the letter of invitation or other notices, tenderers will frequently be required to attend a site showround to be conducted at the site. At the appointed date and time, it is common for the Architect to conduct the site showround at the site to brief the tenderers of the project.

Collection of Tender Documents

After attending the site showround, tenderers will be required to pay the tender deposit and cost of the tender documents prior to collection of the Tender Documents and drawings.

Preparation of tenders

Upon collection of the Tender Documents and drawings, the tenderers will prepare the tender. In a traditional DBB tender (i.e. the tenderer is only required to offer a price which the tenderer wants to be paid for the construction and completion of the project, based on design and drawings provided by the Consultants), the tenderer will ascertain the cost of construction through estimation of quantities of work, sub-contracting and other costs. Thereafter, the tenderer will add a margin for profit. Naturally, the price offered (also known as "Tender Sum") should be as competitively low as possible in order to be successful in the tender. It would be expected by tenderers that the Owner would usually award the project to the lowest tenderer who complies with the tender documents.

The period for tendering may range from a couple of weeks to a few months, depending on the size and complexity of the project.

Tenderers will be required to submit their tenders by the stipulated time on the *Tender Closing Date*. The terms usually provide that late submission will be disqualified and the tender deposit forfeited.

Tenderers will be required to submit their *Form of Tender* by placing it in a tender box (See Fig 2.4). The Form of Tender should be completed by the tenderer with the Tender Sum. A sample of a Form of Tender is shown in Fig 2.5.

Fig. 2.3   (*Continued*)

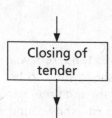

Closing of
tender

Fig. 2.4   Tender box.

There should be strict rules for the Tender Closing procedure to ensure fairness and integrity in the tendering system. For example:

- Tenderers must comply with the deadline for submission of tender.
- At the stipulated time for Tender Closing, the tender box would be closed and no tender would be accepted.
- After tender has closed, a Tender Closing Committee, usually consisting of three persons from different departments (independent witnesses), will open the tender box to record and collect the tenders submitted.

Fig. 2.3 (*Continued*)

FORM OF TENDER

To: ABC Realty Pte. Ltd.
101 High Street
Singapore 123456

Having inspected the site and examined the Drawings, Conditions of Contract, Specifications and other requirements in the Tender Documents in respect of the "Construction and Completion for a 5-storey building at 8 Makepeace Road, Singapore", I/we offer to construct, complete and maintain the whole of the Works comprised in the Tender Documents for a lump sum price of:

(Singapore Dollars: _____

_____ (S$_____)

I/We undertake, if my/our tender is accepted, to complete and deliver the whole of the Works within twelve (12) calendar months from the date of commencement.

I /We agree to the Conditions as follows:

(1) The Tender Documents (ref: TD/001/2016) consisting of the following documents:
   (a) This Form of Tender and the Conditions herein;
   (b) Articles and Conditions of Building Contract, Singapore Institute of Architects, Lump Sum Contract 9th Edition;
   (c) Performance Bond;
   (d) Specifications;
   (e) Breakdown of Cost of Works;
   (f) Drawings
       and other requirement forming part of the Tender Document
       (ref: TD/001/2016)
(2) I/We shall comply with all the terms, conditions and other requirements in the Tender Document.
(3) [other conditions etc...]

_____
Name: _____
Designation: _____
For and on behalf of _____(Name of company)
Date: _____

Fig. 2.5 Form of Tender.

Fig. 2.3 (*Continued*)

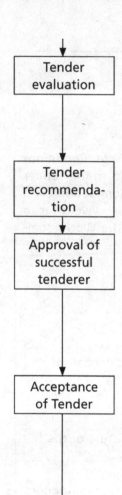

Tender evaluation

Tender evaluation is usually carried out by the Consultants (e.g. Architect, QS). The main criteria assessed are:

- Tender Sum;
- Compliance with Tender Documents.

Tender recommendation

For the sake of fairness, the Consultants usually recommend the lowest tenderer who fully complies with the Tender Documents to the Owner for approval.

Approval of successful tenderer

The Owner approves the recommendation. In respect of corporations and Government tenders, different committees consisting of senior officials are empowered to approve successful tenderers. The committee's approval ensures that no single person decides in the tender approval, thereby providing a system of checks and balance within the approval of tender process.

Acceptance of Tender

Upon approval by the Owner, the Letter of Acceptance or award is posted to the Successful Tenderer. A binding contract takes effect between the Owner and Successful Tenderer upon posting of the Letter of Acceptance.

Upon receipt of the Letter of Acceptance, the Successful Tenderer should get ready to take possession of the Site from the Owner to commence works. Prior to taking over the site, the contractor (successful tenderer) has to ensure that the following are effected in accordance with the contract and copies provided to the Consultants:

- Insurance;
- Security deposit or bank guarantee (or equivalent);
- Construction program (bar chart etc.);
- Other requirements in the contract documents.

Fig. 2.3 (*Continued*)

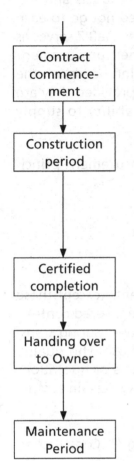

Construction period

The Contractor must complete the whole of the works by the Contract Completion Date or Extended Date (extension of time may be granted to the contractor for reasons expressed in the conditions of contract). If the contractor fails to do so, he may be liable to pay liquidated damages to the Owner.

Certified completion

The works are deemed completed only when the Architect certifies that the works are completed in accordance with the contract.

Handing over to Owner

Upon completion and issue of Temporary Occupation Permit ("TOP") by the relevant Authority, the site is handed over to the Owner by the Contractor. The Owner may take possession of and use the premises.

Maintenance Period

However, the contractor remains liable to repair defects and complete any outstanding works during the Maintenance Period (sometimes known as Defects Liability Period).

2.3.3 GeBIZ

The Singapore Government has developed and set in operation a website portal called GeBIZ[8] for the purpose of calling of tenders and quotations for Government procurement of goods and services. It acts as a one-stop website location for suppliers and contractors to have access to all public sector procurement requirements.

GeBIZ is an electronic platform for purchasers of goods and services to meet the corresponding sellers. The Government departments, statutory boards and other Government agencies ("Government

---

[8] https://www.gebiz.gov.sg/

Entities") need not go to the market to search for products and services which they need. Suppliers and contractors need not go to each Government Entity to market their goods and services. GeBIZ serves as a platform where all Government Entities state their procurement requirements (including development and construction needs) on the GeBIZ portal. Contractors and suppliers may access, provided they are registered, the website, and make known their ability to supply accordingly.

Such a platform supports the fundamental procurement principles of:

- Transparency;
- Fair and Open Competition;
- Value for Money.

All Government Entities' procurement requirements (except those of a confidential nature) are listed on GeBIZ for all registered contractors and suppliers' viewing and tendering access, thus ensuring transparency and fairness. It is *Open Tendering* as all those who qualify may, depending on the situation, submit their tenders or quotations online. Frequently, a Government Entity may specify certain restrictions to tenderers, for example:

- tenderers must be registered with the Building & Construction Authority ("BCA");
- only tenderers registered with specified BCA tendering limit may tender;
- only tenderers registered with specified BCA types of works may tender.

In registering contractors for public sector works, BCA classifies contractors according to their past experience, track record, type of works and financial capacity. Hence, Government Entities would specify their requirement for a contractor according to BCA's classification. In a way, the specification of requirements spells out the pre-qualification of tenderers prior to tendering. It is necessary to ensure the suitability of the tenderers according to their track record and past experience in the type and value of projects that they have successfully completed.

After evaluation, the successful tenderer may be announced online.

GeBIZ is an efficient electronic method for:

- inviting and notifying of Government quotations and tenders,
- providing tenderers with the terms, conditions, specifications and drawings, and other information in respect of a project requirement,
- submitting quotations and tenders by tenderers, and
- announcing the successful tenderers.

As the nature of GeBIZ is *Open Tendering*, there is potential for high competition depending on the number of tenders submitted. There is also good accountability for the funds expended as the tender has been open to all eligible tenderers and the best value for money tender may potentially be accepted.

2.3.4 Financial Limits in Corporate Procurement

In the industry, deciding which of the three methods of tendering for a particular procurement may be part of the corporate governance of the company. It is common for a company to have financial govern-ance, guidelines or procedures for calling tenders or quotations. The Management or the Board of a company may have stipulated financial limits for different tendering methods.

For example, a company may require their staff to comply with the procedures in procurement of goods or services, as shown in Fig. 2.6.

	Value of goods or services to be purchased	Method of procurement
1.	Up to $5,000.00	Direct purchase from supplier or contractor. No need to invite quotations nor call for tender.
2.	Exceeding $5,000.00 but not exceeding $20,000.00	Invite at least three suppliers or contractors to quote.
3.	Exceeding $20,000.00	Call for tenders from at least five tenderers.

Fig. 2.6   Table on Value of goods or services and method of procurement.

The above financial limits provide certainty for the staff of a company to carry out purchasing duties. Without the financial limits, some might invite quotations for the smallest of purchases and some might call a single contractor on the phone to provide renovation works worth hundreds of thousands of dollars.

Small value purchases may be obtained by direct purchase without any need for inviting quotations nor calling of tenders, as the time, costs, and resources expended in the process do not justify the small value purchases. As the value for purchases increases to, say, between $5,000 and $20,000, three independent quotations may be required to ascertain the reasonable market value of the goods or services. A sample letter quotation is shown in Fig. 2.7.

2.3.5 Strengths and Weaknesses of the Three Methods of Tendering

The following table in Fig. 2.8 shows the strengths and weaknesses of the three tendering methods. It would be useful to give some thought to these factors before deciding on a particular tendering method.

From the above table, one would understand that there is no perfect tender method for a particular project. In choosing a particular method of tender, there will be a balancing exercise of what factors are more important which in turn affects the choice for a specific tender method. Naturally, each method has its merits and shortfalls.

Contractor's Letterhead

1 February 2016

ABC Realty Pte. Ltd.
101 High Street
Singapore 654321

Attention: Mr. Tan Seng Huat

Dear Sir

QUOTATION FOR ELECTRICAL WORKS AT FACTORY BLK A, GEYLANG BAHRU, SINGAPORE

As requested, our quotation for electrical works is as follows:

S/No.	Description	Qty	Rate	Amount
1.	Labor for removing 100A metal isolator.	2 hours	$50 per hr.	$100.00
2.	Supply and install power points c/w 2 x 13A switch socket outlet, flush molded box, in PVC trunking/conduit, 6x1 core 2.5mm² PVC cable of 40m run including ring circuit.	30 no.	$130 per no.	$3,900.00
3.	Supply and install 1x4 sq. m PVC insulated cable.	800m	$1.00 per m.	$800.00
4.	Supply labor and material for PVC trunking 50mm x 25mm and making good concrete, brickwork, plastering.	100m	$5.00 per m.	$500.00
			Sub-Total	$5,300.00
			Add: GST	$371.00
			Total	5,671.00

Rgd
Lee Ah Seng
Manager

Fig. 2.7   Sample letter quotation.

	Factors	Open	Selective	Limited	Remarks
1.	Highly competitive in price	Yes	Yes	No	Where many tenderers are tendering as in the Open and Selective Tender, the tender price is expected to be highly competitive and vice-versa.
2.	Fairness to all potential tenderers	Yes	No	No	Where the Owner or Consultants select names of tenderers to tender, those not selected will criticize the process as unfair to them. In Open Tendering, all eligible tenderers may tender.
3.	High integrity in the system of appointing a successful tenderer	Yes	No	No	Once there is human intervention in Selective or Limited Tendering, the integrity of the system of appointing a successful tenderer is in question. In Open tendering, apart from setting the basic criteria, there is no human selection and the success of a tenderer depends on merit.
4.	All tenderers tendering are suitable for the project	No	Yes	Yes	In Open Tendering, many tenderers may satisfy the basic criteria but may still be unsuitable. In Selective and Limited Tender, the tenderers are more suitable as Consultants select those tenderers who have good past performance records of the specific type of work to undertake the project.

No.		No	Yes	Yes	
5.	Allegations of biasness and favoritism	No	Yes	Yes	Naturally, in Selective and Limited Tender, tenderers are so called 'hand-picked' and therefore are open to allegation of biasness and favoritism.
6.	Tenderers required to participate in design prior to tendering	No	Yes	Yes	If tenderers are encouraged to participate in the design process prior to tendering, it would not be practical to adopt Open Tender as there may be too many tenderers in the Open Tender method. In contrast, it may be practical in Selective and Limited Tendering due to smaller number of tenderers.
7.	Longer time for tendering by tenderer(s)	No	No	Yes	In Open and Selective Tender, it is common to give a deadline for the submission of tender. Hence, Consultants decide on the length of time to be given to the tenderers to tender. In Limited Tender, where there is usually only one tenderer, parties negotiate on the price, terms and conditions. This invariably leads to a longer time before a deal is agreed to.

Fig. 2.8 The strengths and weaknesses of the three tender methods.

CHAPTER 3

Project Delivery Methods

In this chapter, we will discuss the following:

3.1 Meaning of Project Delivery Method
3.2 Different Types of Project Delivery Methods
3.3 Risk assessment for Different Payment Models

3.1 Meaning of Project Delivery Method

In the context of construction work, the *Project Delivery Method* is a system for organizing the finance, design, construction, operation and maintenance service for a building or facility by entering into contracts with various parties.

3.2 Different Types of Project Delivery Methods

Traditionally, the Owner employs a Contractor to construct and complete a building or structure. The Contractor has no part in the design of the building. The Owner engages the Consultants to provide the design and specifications. Both the Contractor and Consultants will be paid by the Owner. The Contractor's duty is to construct according to the Consultant's design and specifications. Upon completion, the Contractor will deliver the project to the Owner, who will himself provide the maintenance of the building or get other service providers to maintain it. This method of project delivery is called the Design-bid-build method.

Through needs and innovation, various other Project Delivery Methods have emerged in the industry. Many of the more innovative methods involved Contractors in the design, finance, operation and

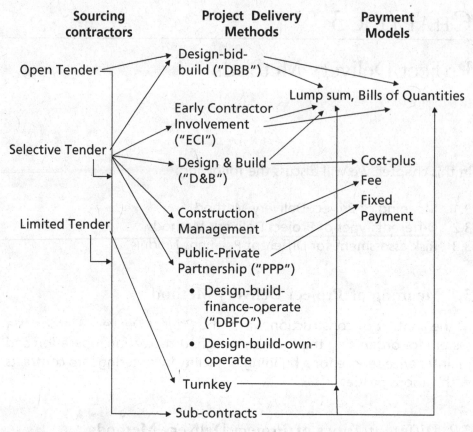

Fig. 3.1   Sourcing contractors, project delivery methods and payment model.

maintenance as well. For ease of explanation, please refer to Fig. 3.1 for some of the different types of Project Delivery Methods.

Fig. 3.1 shows the different types of Project Delivery Methods in connection with the types of sourcing contractors and payment models. The arrows show the more common connection, e.g. it is more common to have Open Tender connected to DBB than Open Tender connected to ECI due to practical constraints of having all tenderers in Open Tender taking part in ECI. But, if the number of tenderers in Open Tendering is manageable, it may be possible to connect Open Tender to ECI. Hence, in understanding Fig. 3.1, one must not dismiss the possibility of a connection in the figure above where there is no arrow. Fig. 3.1 should not be accepted rigidly but with flexibility for change due to different circumstances.

We will now discuss some of the Project Delivery Methods.

3.2.1 Design-bid-build ("DBB")

3.2.1.1 What is DBB?

As mentioned, this is the traditional method and most of the projects in Singapore make use of this method. The Owner engages Consultants to design the project and draft tender documents. Then, there is calling of tender based on tender documents drafted (including the design). Tenderers will bid (tender) for the project. The Owner employs the successful tenderer to construct and complete the building or facility. The Contractor has to carry out the construction in accordance with the design and specifications of the Consultants. Upon completion, the Contractor delivers the project to the Owner. See Fig. 3.2 for DBB method.

The flowchart for the DBB method may be viewed in Fig 2.3.

3.2.1.2 Why DBB?

Owners engage Consultants separately from the Contractor as Consultants are professionals experienced in design and management in their area of expertise. In the past, Contractors had little

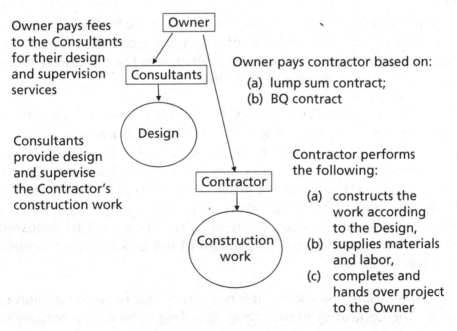

Fig. 3.2   Model for DBB method.

professional training, but today many Contractors are not only highly skilled but also have the support of in-house Architects, Engineers and other professionals.

Due partly to tradition, Owners would engage Consultants to design, manage and supervise the construction work. This also allows Owners to have direct control over the Consultants in the design and management of the construction work. In a way, the Consultants are the Owner's agents to make sure that works carried out by the Contractors are up to speed and of good quality. The respective work of the Consultants and Contractors forms the traditional divide between them.

3.2.1.3 Payment to contractor under DBB

Under the DBB method, payments to a Contractor are usually made under a Lump Sum contract or Bills of Quantities ("BQ") contract (see Fig. 3.1). A Lump Sum contract is more common than a BQ contract.

In a Lump Sum contract, the Contractor tenders a sum for the construction and completion of the project based on drawings and specifications. Subject to any variations requested by the Consultants, the Contractor will be paid the lump sum tendered.

In a BQ contract, the Contractor tenders on the basis of the bills of quantities, etc. Upon completion of works, the works are re-measured according to as-built drawings and payments to the Contractor made according to the re-measured quantities.

Hence, in a Lump Sum contract, the Contractor has to ascertain the quantities of work for himself. Any under-measurement or other errors in quantifying the works during preparation of tender will not be the grounds for compensation by the Owner. In a BQ contract, the Owner's QS will measure and provide the bills of quantities in the Tender Documents for the Contractor's pricing during preparation of tender. Any under-measurement or other errors will be adjusted through the process of re-measurement of the as-built drawings upon completion of works.

Frequently, re-measurement is not carried out upon completion of works. The Contractor in such cases, is willing to be paid according to the original BQ prepared by the QS. The advantage to the Contractor in not requiring re-measurement is the earlier completion of the Final Account and final payment to him.

3.2.1.4 Examples of DBB projects

Examples of projects using DBB are many HDB housing projects, developers' condominiums and apartments, office buildings and other common facilities and structures. Some more elaborate and decorated buildings may also make use of DBB, e.g. hotels.

3.2.1.5 Sourcing contractors

The Contractors in DBB projects may be sourced through Open Tender or Selective Tender (shown in Fig. 3.1). In exceptional circumstances, a Contractor may be sourced from Limited Tender to undertake the project through DBB.

3.2.1.6 Risks associated with DBB

Risks associated with the DBB method are shown in Fig. 3.3.

3.2.2 Early Contractor Involvement ("ECI")

3.2.2.1 What is ECI?

ECI is a project delivery method that allows tenderers' early involvement and feedback prior to acceptance of tender based on the following.

- Buildability of the design;
- Time schedule and plans for the project;
- Value engineering;
- Value management;
- Saving in time, cost and resources in construction due to the application of ECI.

The ECI process may be represented by the following flow chart in Fig 3.4.

	Types of risks	Risk faced by Owner	Risk faced by Contractor	Remarks
1.	Payment for the Contract Sum	Yes	No	If the Contractor is a Main Contractor with Sub-Contractors, then the Main Contractor assumes risks in payment to the Sub-Contractors.
2.	Defective design	Yes	No	Owner may recover against the Consultant who provides the defective design.
3.	Construction and completion of building	No	Yes	
4.	Increase in material price	No	Yes	If there is a material price fluctuation clause, then the Contractor is entitled to claim for the increase in material price.
5.	Increase in labor price	No	Yes	
6.	Delay by workmen and subcontractors	No	Yes	Contractor will be liable to Liquidated Damages for delay beyond contractual Extension Of Time.
7.	Any work which the Contractor fails to measure and price in the Contract	No	Yes	In Lump Sum contract, the Owner does not bear the risk of Contractor's error in measurement and pricing.

Fig. 3.3   Risks associated with DBB method.

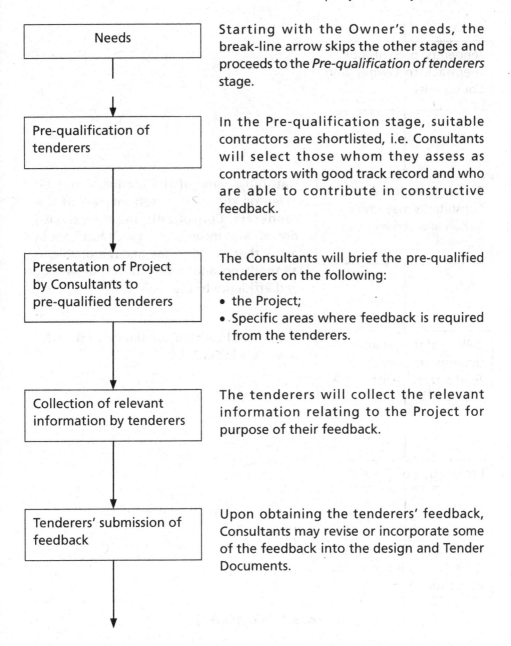

Starting with the Owner's needs, the break-line arrow skips the other stages and proceeds to the *Pre-qualification of tenderers* stage.

In the Pre-qualification stage, suitable contractors are shortlisted, i.e. Consultants will select those whom they assess as contractors with good track record and who are able to contribute in constructive feedback.

The Consultants will brief the pre-qualified tenderers on the following:

- the Project;
- Specific areas where feedback is required from the tenderers.

The tenderers will collect the relevant information relating to the Project for purpose of their feedback.

Upon obtaining the tenderers' feedback, Consultants may revise or incorporate some of the feedback into the design and Tender Documents.

Fig. 3.4   Flowchart for ECI method.

Tenderers' presentation of feedback to Owner and Consultants	
Based on feedback, Consultants may revise design and Tender Documents	Naturally, some of the feedback may be given on the basis of self-interest of the tenderers. Consultants should exercise discretion in incorporating such feedback in a way that would be fair and reasonable to all tenderers and in the interest of savings and efficiency to the project.
Calling of tender and tenderers to collect Tender Documents	The stages hereafter are similar to the DBB flowchart in Fig 2.3.
Preparation of Tenders	
Other activities to follow as in DBB	

Fig. 3.4 *(Continued)*

Fig. 3.4 shows a possible process for ECI. Naturally, one could have the ECI process much later during the Preparation of Tenders stage or earlier, depending on the circumstances of each project. Activities within the ECI process may be varied for more consultations, if necessary. The objective of the exercise is to solicit feedback from the Tenderers in order to improve the Design and Tender Documents in the interest of the Project.

3.2.2.2 Why ECI?

Being experienced in construction, Contractors should know the practical aspects of constructing a certain design and could advise on its buildability. Some designs may be complicated and difficult to build on the construction site and Contractors could make this known to Consultants prior to acceptance of tender so that necessary changes may be made. Similarly by virtue of their experience, Contractors may provide constructive feedback on time schedule and plans for the project, more efficient engineering techniques to achieve a certain result, etc.

Hence, there are good reasons to involve the Contractors early during the conceptual plan, schematic design or even preparation of tender stage in soliciting their feedback and revising the plans, drawings and other terms and conditions arising from such feedback.

3.2.2.3 Payment to contractor under ECI

Payment to Contractors under the ECI process may be under a Lump Sum contract or BQ contract (see Fig. 3.1).

As you would note, there is little difference between the ECI and the process of DBB except that contractors are brought in at an earlier stage for the soliciting of feedback. After feedback is given, the process continues as in DBB process with the collection, preparation and submission of tender, etc.

3.2.2.4 Sourcing contractors

Sourcing Contractors may be through Selective Tender (see Fig. 3.1). Given the larger number of tenderers in Open Tender, it may not be practical to carry out the ECI process.

The number of Contractors to take part in an ECI process may be from four to six Contractors.

3.2.2.5 Risks associated with ECI

The risks associated with the ECI process are similar to those of the DBB process, as shown in Fig. 3.3. It is submitted that Consultants may not abdicate their responsibility by blaming the Contractors for their feedback. Ultimately, it is for the Consultants to decide and within their scope of duty of care whether to incorporate the Contractors' feedback into the design and Tender Documents. In the Author's view, the Consultants should not be held responsible if any loss and damage results from the Owner insisting on incorporating the Contractors' feedback, against the Consultants' advice.

One weakness in the ECI process is the prolonging of the pre-contract period. The ECI process is estimated to prolong the pre-contract period by one to three months. The Owner and Consultants should take this into consideration in the overall scheduling for the project.

Other risks may be associated with the feedback from Contractors. Some feedback may be to the advantage of the Contractor who suggests a particular change that is disadvantageous to other Contractors. Incorporating such a feedback into the design and tender documents would create a *non-level playing field* where other Contractors would face greater difficulties in fulfilling the requirement incorporated during the tender stage. Hence, it is suggested that Consultants should be vigilant and transparent, disclosing changes suggested during the feedback process to all Contractors and requesting for their further feedback before incorporating these feedback into the Tender Documents.

Thus, the ECI process in Fig. 3.4 may be lengthened due to more consultative activities in the interest of fairness and transparency. Consequently, higher costs and resources may be expended in the process.

3.2.3 Design & Build ("D&B")

3.2.3.1 What is D&B?

D&B is a project delivery method where tenderers provide not only construction services but also the design for the project. In other words, the Owner does not engage Consultants but relies on the Contractor to provide the design and build as well. In order to provide the design service to the Owner, the Contractor may employ in-house architects, engineers and other professionals or they may outsource the design work to Consultants (i.e. the Contractor engages external Consultants to provide the design).

Contractually, the Owner engages the Contractor to design and build the project. If the Contractor outsources the design work, he remains liable to the Owner for the design. If the design is defective, resulting in loss and damage, the Owner may take action against the Contractor for breach of contract. Consequently, the Contractor may take action against the outsourced Consultants for their defective design for breach of contract under the contract between the Contractor and Consultant.

A flowchart of the D&B method is presented in Fig. 3.5 below.

Subject to variations, the flowchart for the D&B method is similar to the DBB method in Fig 2.3, except that there is no appointment of Consultants and related activities. Tenderers would be pre-qualified based on their experience and track record in undertaking D&B projects. The Tender Documents would require tenderers to provide design and build services for the approval of the Owner. Each tenderer may be required to submit design, plans, costs, time schedule and other information for the Owner's evaluation and decision. With the design and information of each tenderer, the Owner decides on the successful tenderer.

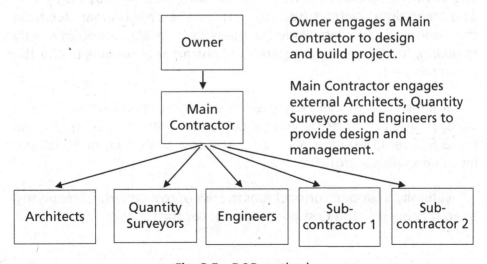

Fig. 3.5   D&B method.

3.2.3.2 Why D&B?

There are many reasons for considering D&B for a project.

- It is a *one-stop shop* for the Owner. The Owner will not need to enter into separate contracts with the Architect, Engineers and other professionals. Consequently, the Owner only looks to and instructs one party, the Main Contractor, for the design and construction of the project. The Owner frees himself of the considerable time and attention devoted to many parties and focuses instead only on the Main Contractor.
- Since the Main Contractor is able to combine both design and construction, the design is expected to be more practical and buildable.
- There may be a reduction of time for completion of the project as the traditional divide in design, tendering and construction no longer exists. Some of the activities separated in the traditional DBB method may run concurrently (e.g. some design and construction activities) or be removed altogether. Furthermore, there is no waiting for Consultants' approval of the design and works.
- There may also be a reduction in the costs of construction due to a more buildable design, value engineering and management by the Main Contractor.

3.2.3.3 Payment to contractor under D&B

The payment to Contractors under the D&B method may be by way of a Lump Sum contract (see Fig. 3.1). As the design emanates from the Contractor, it is less likely for there to be a BQ, although strictly speaking, the Owner may require the Contractor to provide BQ for the project as well.

In some cases of D&B, e.g. specialized high-tech electronic installations, payment may be by way of *Cost-plus contract* (see Fig. 3.1 and Fig. 3.6). The Contractor is paid for the cost of design, material and installation plus a profit.

Naturally, Cost-plus contract runs the risk of cost over-run. Frequently, a cap is applied to the Cost-plus payment contract.

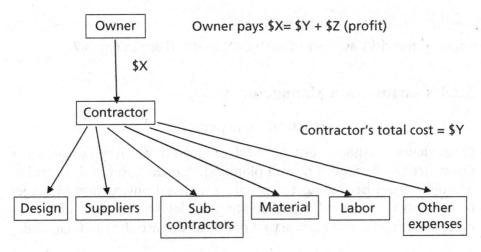

Fig. 3.6   Cost-plus contract model.

3.2.3.4 Examples of D&B

In Singapore, many wafer-fabrication plants make use of the D&B method.

3.2.3.5 Sourcing Contractors

Generally, the D&B method makes use of the Selective Tender method (see Fig. 3.1) to source for Contractors. An Open Tender sourcing may result in too many tenderers. The Selective Tender sourcing with four to six tenderers to provide proposals for design would be ideal.

In some instances of Selective Tender for D&B, the tenderers are required to tender by submitting two-envelopes, i.e. the *two-envelope bidding system*. In the first envelope, the tenderer will submit his technical information that may consist of design and specifications proposed for the project. In the second envelope, the tenderer will submit his price. Theoretically, the first envelopes to all tenderers should be opened first for evaluation of the technical information. Only those whose technical information are acceptable would be subject to the opening up of their second envelopes to assess their price. Naturally, the successful tenderer would be one whose technical information is acceptable when the first envelopes are opened and whose price is competitive in the opening of the second envelopes.

3.2.3.6 Risks associated with D&B

Some of the risks associated with D&B method are in Fig. 3.7.

3.2.4 Construction Management

3.2.4.1 What is construction management?

Construction management method of project delivery requires the Construction Manager's overall planning, coordination and control of a project from beginning to completion. The Construction Manager co-ordinates all the different building trades to construct and complete the project. The contractual relationships are shown in Fig. 3.8.

Instead of engaging a Contractor to construct and complete the whole of the works, the Owner engages a Construction Manager to undertake the overall planning, co-ordination and control of the works. In Fig. 3.8, the Owner enters into contract with the Architect, Engineers and QS in respect of the consultancy services. The Owner also enters into contract with the Construction Manager, Trade Contractor 1, e.g. earthworks and excavation contractor, Trade Contractor 2, e.g. contractor for the supply and placing of concrete, and other trade contractors.

The contractual relationships are shown by the arrow lines. The Construction Manager plans, co-ordinates and controls the trade contractors, as shown in straight lines (without arrow heads), to ensure that the whole of the works is completed.

In order for the Construction Manager to plan and co-ordinate the trade contractors, this person should be engaged and brought into the project team early, i.e. during the design stage of the project. With the completion of the design, the Construction Manager would plan, schedule and co-ordinate the works for the trade contractors.

3.2.4.2 Why construction management?

In Singapore, it is uncommon to employ a Construction Manager for a project. The management of the works on site is usually undertaken by the Main Contractor. The Main Contractor co-ordinates, plans and controls the sub-contractors and the whole of the works. But in large projects, the construction management project delivery method may be used to achieve specialization in the planning and management of works.

	Types of risks	Risk faced by Owner	Risk faced by Contractor	Remarks
1.	Payment for the Contract Sum	Yes	No	
2.	Defective design	No	Yes	
3.	Construction and completion of building	No	Yes	
4.	Increase in material price	No	Yes	If there is a material price fluctuation clause, then the Contractor is entitled to claim for the increase in material price.
5.	Increase in labor price	No	Yes	
6.	Delay by workmen and subcontractors	No	Yes	Contractor will be liable to Liquidated Damages for delay beyond contractual Extension of Time.
7.	Any work which the Contractor fails to measure and price in the Contract	No	Yes	In Lump Sum contract.

Fig. 3.7   Risks associated with D&B method.

	Types of risks	Risk faced by Owner	Risk faced by Contractor	Remarks
8.	Owner has less control over the Architect and professionals in the design of the project	Yes	No	The Owner bears the risk of less control over the Architect and professionals as the Architect and professionals are engaged by the Contractor (not Owner).
9.	Less accountability of Architect and professionals to the Owner	Yes	No	The Owner bears the risk of less accountability from the Architect and professionals as they are engaged by the Contractor (not Owner).
10.	Architect and professionals act in the interest of the Contractor and less so in the interest of the Owner	Yes	No	The Owner bears the risk that the Architect and professionals act in the interest of the Contractor (not Owner) due to their obligations to the Contractor under their contract of engagement.
11.	As the design emanates from the contractor, strict compliance with the design may be difficult as the Architect and professionals act in the interest of the Contractor.	Yes	No	If the Architect and professionals owe their duties to the Contractor, it is less likely for them to insist on strict compliance with their design.

Fig. 3.7   *(Continued)*

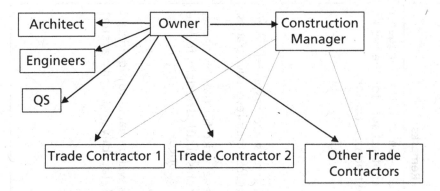

Fig. 3.8   Model of Construction Management method.

3.2.4.3 Payment to the construction manager

A Construction Manager is engaged on a fixed fee basis (see Fig. 3.1), much like the other Consultants.

3.2.4.4 Examples of construction management projects

An example of construction management services provided is part of the Marina Bay Sands project.

3.2.4.5 Sourcing construction managers

Construction Managers would be sourced in similar manner as that for Architects and other consultants through a selective process (see Fig. 3.1). The evaluation is based on their track record, services to be provided and the fee quoted.

3.2.4.6 Risk associated with the construction management method

The risks associated with the Construction Management method are in Fig. 3.9.

3.2.5 Public-Private Partnership ("PPP")

3.2.5.1 What is PPP?

Traditionally, the public sector engages Contractors from the private sector to construct facilities or supply equipment. Upon completion,

	Types of risks	Risk faced by Owner	Risk faced by Construction Manager	Remarks
1.	Payment of the Construction Management fee	Yes	No	The Owner pays the Construction Manager a fee and also the contract sum for the works carried out by the trade contractors.
2.	Payment to the trade contractors	Yes	No	
3.	Defective design	Yes	No	The trade contractors are responsible to construct and complete their respective trades.
4.	Construction and completion	Yes	No	
5.	Increase in material price	No	No	If there is a material price fluctuation clause, then the Contractor is entitled to claim for the increase in material price.
6.	Increase in labor price	No	No	
7.	Delay by workmen and subcontractors	Yes	No	Trade contractors will be liable to Liquidated Damages for delay beyond contractual Extension of Time.
8.	Any work which the Trade Contractors fails to measure and price in the Contract	No	No	

Fig. 3.9   Risks associated with Construction Management method.

the public agencies will then operate the facilities or equipment. Public-Private Partnership ("PPP") is an alternative form of project delivery method where the public sector focuses on acquiring services at the most cost-effective basis rather than owning and operating facilities. It requires a long-term partnering relationship between the public and private sectors to deliver services.

The comparison of timelines between the traditional and PPP project delivery methods is in Fig. 3.10.

In the Traditional project delivery method (e.g. DBB), different parties provide and manage the project timeline as shown in the Traditional method timeline. In the PPP project delivery method, only the PPP service provider provides and manages the timeline right through. There is a reduction in time, cost and resources without changes in parties.

As could be seen in Fig. 3.10, a PPP service provider provides the design, construction, operation and management of the facility. In addition, the PPP service provider also obtains his own financing for the whole project (whether through bank loans or funds from investors). Such financing may be large as the project, in most cases, is large and costly. In order to be motivated, the PPP service provider is usually granted a lease of the facility for a number of years. During the lease

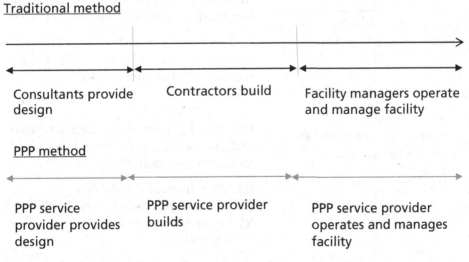

Fig. 3.10   Comparison of timelines between Traditional and PPP project delivery methods.

period, the PPP service provider will operate and manage the facility as a business concern to recover his substantial investment in the project.

As the name suggests, PPP signifies the partnership between Public and Private sectors. In more precise terms, the above method in PPP is known as the Design-Build-Finance-Operate (DBFO) model. DBFO is the most common model of PPP in Singapore. Another model making use of PPP method is the Design-Build-Operate model (DBO).

A brief flow-chart for DBFO may be represented in Fig. 3.11.

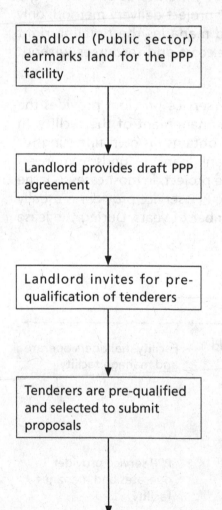

Landlord (Public sector) earmarks land for the PPP facility	The public sector may have to acquire land for the PPP facility.
Landlord provides draft PPP agreement	The PPP agreement may consist of: (a) Building Agreement; (b) Lease. The successful PPP service provider may have to purchase the land from the Landlord, usually for a lease of a period of years.
Landlord invites for pre-qualification of tenderers	Announcement would be made worldwide for interested parties to submit pre-qualification details.
Tenderers are pre-qualified and selected to submit proposals	Interested parties would be shortlisted and selected tenderers would be invited to submit proposals (RFP): (a) Mock-up model of facility; (b) Financial payment for the land; (c) Payment for operating the Facility or otherwise.

Fig. 3.11   Flowchart for DBFO.

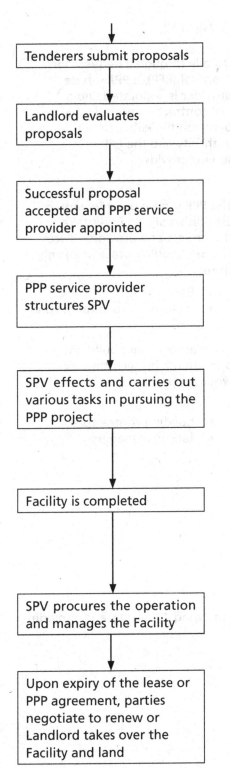

Upon acceptance of proposal, the PPP service provider structures the Special Purpose Vehicle ("SPV").

The SPV effects the following:
(a) Pay the purchase price of the land to the Landlord;
(b) Obtain bank loans and other forms of Financing;

(c) Appoint design consultants;
(d) Develop and complete design;
(e) Procure construction contractors, interior designers, etc.;
(f) Engage facility managers.

Fig. 3.11   (*Continued*)

A structure for the DBFO model is shown in Fig. 3.12.

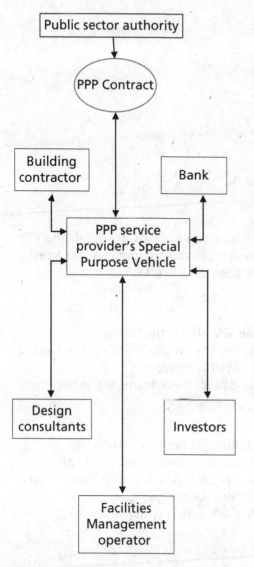

Usually through a Request for Proposal (RFP), a PPP service provider is appointed and a PPP contract is entered into between the Public sector authority and the PPP service provider.

The PPP service provider would establish a SPV to undertake the project and provide services. The SPV would obtain financing from:

- Bank loans;
- Investors' provision of funds.

In order to design, build and operate the project, the SPV would appoint:

- design consultants,
- building contractors
- facility managers

Fig. 3.12   Structure of DBFO model for obtaining financing and providing services.

3.2.5.2 Why PPP?

There are many reasons for the PPP method of project delivery, some of which are as follows:

- The Public Sector may not have the knowledge and experience in providing a particular service, whereas the private sector has the specialized knowledge and experience;
- Risk is transferred from the Public Sector to the Private Sector as the Private Sector assumes the risk of design, build, finance and operation of the facility;
- The Private Sector strives to ensure success of the project as a PPP service provider is motivated to maximize profit through the success of the project;
- Competitive proposals by a number of Private Sector organizations give rise to the best design and service;
- The Public Sector's capital expenditure for the project is minimal as the financial expenditure and risks have been transferred to the PPP service provider.

3.2.5.3 Payment (income) to the PPP service provider

As may be seen from Fig. 3.13, the Public Sector Authority provides an Annual Payment to the PPP service provider (see Fig. 3.1), usually as long as the Facility is available for use. Other income may depend on the business of the PPP service provider in having events, food & beverage tenants in the premises, etc.

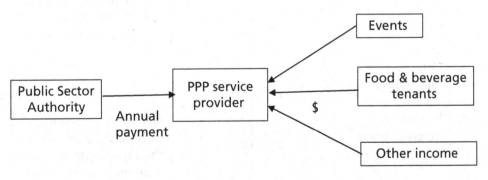

Fig. 3.13   Sources of income to the PPP service provider.

3.2.5.4 Examples of PPP projects

A well-known example of a PPP project is the Singapore Sports Hub. Other PPP projects include ITE College West and the Tuas Desalination Plant.

3.2.5.5 Sourcing PPP service providers

A common method is the Selective Tender (see Fig. 3.1) through RFP (Request for proposals). The flow process has been shown in Fig. 3.11.

3.2.5.6 Risks associated with PPP

Some of the risks associated with PPP are in Fig. 3.14.

   As seen in Fig. 3.14, most of the risks are borne by the PPP service provider. The disadvantage to the Public sector authority may be the loss of control and management of the facility.

	Types of risks	Risk faced by Govt Sector Authority	Risk faced by PPP service provider	Remarks
1.	Payment to the Contractors	No	Yes	In the DBFO model, the PPP service provider has to finance the construction works.
2.	Defective design	No	Yes	The PPP service provider may take action against the design consultants for defective design.
3.	Construction and completion	No	Yes	
4.	Increase in material price	No	No	If there is a material price fluctuation clause, then the Contractor is entitled to claim for the increase in material price.

Fig. 3.14   Risks associated with PPP model.

	Types of risks	Risk faced by Govt Sector Authority	Risk faced by PPP service provider	Remarks
5.	Increase in labor price	No	No	
6.	Delay by workmen and subcontractors	No	Yes	Contractors will be liable to pay Liquidated Damages to the PPP service provider for delay beyond contractual Extension of Time.
7.	Any work which the Contractors fail to measure and price in the Contract	No	No	In a Lump Sum contract, the Contractors assume such risks.
8.	Loss of possession and control over the land and facility	Yes	No	
9.	Loss of control over the management and operation of the facility	Yes	No	

Fig. 3.14   (*Continued*)

3.2.6 Turnkey

3.2.6.1 What is Turnkey?

A Turnkey project is one where the Owner engages a Contractor to design, build and hand over a project to the Owner with little or no supervision from the Owner or the Consultants (see Fig. 3.15). Thus the Owner's involvement in the design and construction is very minimal.

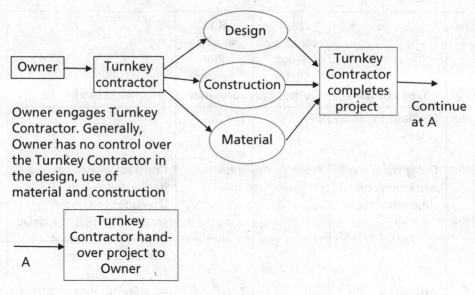

Fig. 3.15   Model of Turnkey method.

When completed, the 'key' to the facility is handed over, and the Owner, as it were, 'turns the key' to open the facility.

Naturally, the Owner must initially give some idea of what he wishes to see as the finished product. Then, the Turnkey Contractor, very much on his own, designs, constructs and delivers the finished product to the Owner. This type of project delivery method is uncommon in Singapore.

3.2.6.2 Why Turnkey?

In specialist work where Consultants are not able to design or supervise the work, the Owner very much leaves the design and construction to the Turnkey Contractor, who is a specialist in the design and construction of that particular product.

3.2.6.3 Payment to the Turnkey contractor

Payment to a Turnkey Contractor may be on a lump sum basis (see Fig 3.1) with staged payments over the period of design and construction. Sometimes, payment may be on a Cost-plus basis.

3.2.6.4 Example of Turnkey projects

Examples of Turnkey projects may be some of the rides in amusement parks. Such rides may be specialized work in design and construction. Each ride may be differently conceptualized, planned, designed and constructed.

The Owner may request the specialist Turnkey contractor to conceptualize and provide a new ride that would not be within the capability of Consultants to supervise or monitor.

3.2.6.5 Sourcing Turnkey contractors

Sourcing a Turnkey Contractor is usually by Limited Tender (see Fig. 3.1), as only one Contractor is known to provide the specialist work.

3.2.6.6 Risks associated with Turnkey projects

The risks are in Fig. 3.16.

	Types of risks	Risk faced by Owner	Risk faced by Turnkey Contractor	Remarks
1.	Payment to the Contractor	Yes	No	
2.	Defective design	No	Yes	As the design and construction are not supervised or checked by third parties, there is a risk of defective design or construction.
3.	Construction and completion	No	Yes	

Fig. 3.16 Risks associated with Turnkey projects.

	Types of risks	Risk faced by Owner	Risk faced by Turnkey Contractor	Remarks
4.	Increase in material price	No	Yes	In lump sum contract, the Contractor assumes the risk unless there is a material price fluctuation clause.
5.	Increase in labor price	No	Yes	
6.	Delay by workmen and subcontractors	No	Yes	Turnkey contractor will be liable to pay LD to the Owner for delay beyond contractual Extension Of Time.
7.	Any work which the Contractors fails to measure and price in the Contract	No	Yes	In lump sum contract, the Turkey Contractor assumes the risk.

Fig. 3.16   (*Continued*)

3.2.7 Sub-Contracts

3.2.7.1 What are sub-contracts?

When the Owner engages a Contractor, whether in DBB, D&B or other project delivery methods, the Contractor frequently Sub-Contracts part of his work to Sub-Contractors. Sub-Contracting is a common project delivery method. No single contractor is able to construct and complete the whole building or facility, including the mechanical, electrical ("M&E Works") and other specialist works.

It is common for the Contractor (known as the "Main Contractor") to Sub-Contract the M&E Works and other specialist works to Sub-Contractors. A simple model for Sub-Contracting works is in Fig. 3.17

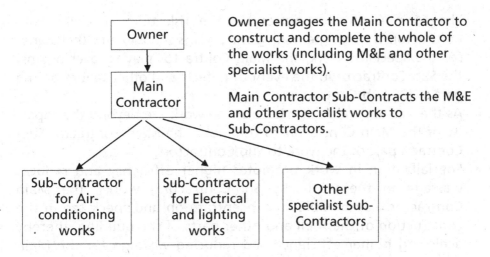

Owner engages the Main Contractor to construct and complete the whole of the works (including M&E and other specialist works).

Main Contractor Sub-Contracts the M&E and other specialist works to Sub-Contractors.

Fig. 3.17   Model for Sub-Contract work.

Where the Main Contractor selects his own Sub-Contractors for the works, these Sub-Contractors are known as Domestic Subcontractors ("DomSC").

It is also common for Owners to select Sub-Contractors. These are known as *Nominated Sub-Contractors* ("NSC"), where the Owner made known to the Main Contractor that the Owner would nominate Sub-Contractors, but the identities of such Sub-Contractors are not made known to the Main Contractor prior to their engagement; or *Designated Sub-Contractors* ("DSC"), where the identities of the Sub-Contractors are made known to the Main Contractor prior to their engagement.

3.2.7.2 Why sub-contract?

Some of the reasons for Sub-Contracting are as follows:

- The Main Contractor may not have the experience and expertise to carry out some of the works, e.g. M&E works;
- Due to their experience, the specialist Sub-Contractors may be able to carry out the works at reduced cost and time compared to the Main Contractor;
- Due to their experience, the specialist Sub-Contractor may be able to carry out better quality work compared to the Main Contractor;

- The Main Contractor outsources the risk to a specialist Sub-Contractor. Though the Main Contractor is still liable to the Owner for the specialist work, the Main Contractor may recover against the Sub-Contractor in the event of defects and other failure on the part of the Sub-Contractor;
- As the size and value of whole of the work are beyond the capacity of the Main Contractor to complete, the Main Contractor Sub-Contracts part of the work to Sub-Contractors;
- Specialization in work promotes higher efficiency and reduces wastage in the work. By Sub-Contracting works, the Main Contractor may be able to concentrate on and specialize in the construction of the shell and other parts of the building, thereby achieving higher efficiency and reducing wastage for the Main Contractor.

3.2.7.3 Payment to sub-contractors

Payment to Sub-Contractors may be under a Lump Sum contract or a BQ contract (see Fig. 3.1).

As there is no contract between the Owner and Sub-Contractors (see Fig. 3.17), Owners may not pay the Sub-Contractors directly for their work. Any payment to the Sub-Contractors must be made through the Main Contractor, whether the Sub-Contractor is a NSC, DSC or DomSC.

Furthermore, the amount of payment to Main Contractor for the work by the NSC or DSC should be notified to the Main Contractor and respective NSC and DSC. With such notification, the Main Contractor should pay the same to the respective NSC or DSC.

In respect of works carried out by DomSC, such works are treated as though they are carried out by the Main Contractor. Any payment for the DomSC's works is paid to the Main Contractor without notification of amount due to the DomSC.

Fig. 3.18 shows the process of payment to the NSC for Air-conditioning and Mechanical ventilation works. Payment to the DSC will follow the same process.

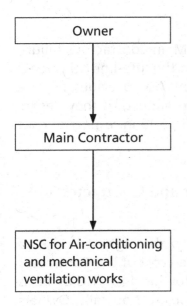

Owner pays Main Contractor for each progress payment and other payments. Progress payments and other payments may include amount payable to NSC. In respect of each progress payment, the Owner's QS would advise the Main Contractor and respective NSC the amount to be paid to the NSC.

Main Contractor should pay the NSC in accordance with the amount advised by the Owner's QS.

Fig. 3.18   Payment to NSC.

3.2.7.4 Example of sub-contracts

As mentioned above, it is common for Sub-Contracts to take place in construction contracts. In particular, M&E works will usually be sub-contracted to specialist mechanical and electrical contractors.

3.2.7.5 Sourcing sub-contractors

Sub-Contractors are usually sourced by tender, similar to the manner in which Main Contractors are sourced. Sourcing Sub-Contractors may be by Open Tender, Selective Tender or Limited Tender (see Fig. 3.1). The process for tendering may follow the flow-chart in Fig 2.3.

Although the Owner's QS and M&E Consultants may be drafting and collating the NSC or DSC tender documents, ultimately the Sub-Contract is entered into between the Main Contractor and NSC or DSC. The Owner has no part in the Sub-Contract. Hence, the NSC or DSC tender documents should make clear that the successful tenderer would be required to enter into a Sub-Contract with the Main Contractor.

3.2.7.6 Risks associated with sub-contracts

Many cases of adjudication arise due to the Main Contractor's failure to pay Sub-Contractors. In order to maintain the 'life-line' payments to Sub-Contractors, the Security of Payment Act[1] provides for the principals and owners (i.e. the Owner of the project), if they desire, to pay directly to the Sub-Contractors. See Fig 3.19 in respect of risks associated with Sub-Contract.

3.3 Risks Assessment Between Owner and Contractor for Different Payment Models

Traditionally, Cost-plus contracts rank among the highest in risk to the Owner. The Owner has to pay the incurred cost of the Contractor for the project plus a profit. The Contractor has low risk as he is assured that his cost would be paid plus a profit. Naturally, Owners will avoid this form of contract due to the inherent risks. However, inroads have been built into the traditional Cost-plus by including a cap to Owner's payment and other incentives to the Contractor to reduce cost.

A BQ contract requires the contractor to price the works according to quantities provided by the Owner's QS. Upon completion of works, there is a re-measurement of quantities according to as-built drawings. Thus, the Owner will pay the Contractor according to the actual quantities constructed. Frequently, parties may agree not to re-measure but pay according to the original BQ.

In Lump Sum contracts, the Owner's QS does not provide quantities for pricing by the Contractor. Upon completion of works, there is no re-measurement of quantities for purpose of payment. The Contractor measures quantities based on tender drawings for the purpose of preparing the tender. He takes the risk of any errors in measurement. He will not be compensated if he under-measures and tenders too low. Compared to Cost-plus and BQ contracts, the Contractor has to bear higher risk in Lump Sum contracts. Lump Sum contracts take less time to draft and collate as the Owner's QS need not measure quantities.

---

[1]Building & Construction Industry Security of Payment Act

	Types of risks	Risk faced by Main Contractor	Risk faced by Sub-Contractor	Remarks
1.	Payment to the Sub-Contractor	Yes	N.A.	Failure of payment by Main Contractor to the Sub-Contractor is not uncommon.
2.	Defective design in Sub-Contract	No	No	Sub-Contractor will not be liable for design if he did not provide the design. If there is loss and damage arising from defective design, the Consultants may ultimately be liable as the party who provided the design.
3.	Construction and completion of Sub-Contract works	No	Yes	
4.	Increase in material price	No	Yes	Unless payment is based on Cost-plus basis, the Sub-contractor will not be compensated for increase in material price.
5.	Increase in labor price	No	Yes	Unless payment is based on Cost-plus basis, the Sub-Contractor will not be compensated for increase in labor price.
6.	Delay due to Sub-Contractor	No	Yes	Sub-Contractor may be liable to pay Liquidated Damages to the Main Contractor for delay beyond contractual Extension Of Time.
7.	Any work which the Contractor fails to measure and price in the Contract	No	Yes	Unless payment is based on Cost-plus, the Sub-Contractor may not be compensated.

Fig. 3.19 Risks associated with Sub-Contract.

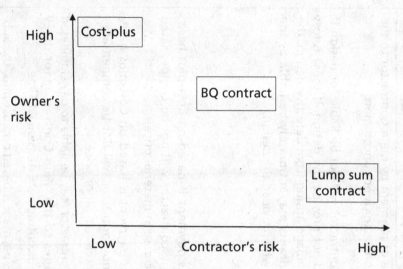

Fig. 3.20 Risk assessment between Owner and Contractor for different payment models.

It is also less costly to the Owner, as the fees for drafting and collating a Lump Sum tender are usually less than those for a BQ tender. Hence, Lump Sum contracts are widely used in Singapore.

Fig. 3.20 shows the level of risk assumed by Owner and Contractor for different types of payment models.

Chapter 4

Contract Document

This chapter will discuss some parts of a Contract Document as follows:

4.1   Conditions of Contract
4.2   Form of Tender
4.3   Form for Insurer Bond/Bank Guarantee
4.4   Roofing and other Guarantees or Warranties
4.5   Preliminaries
4.6   PC (Prime Cost) & Provisional Sum
4.7   Particular Specification
4.8   PC (Prime Cost) Supply Rate
4.9   General Specifications
4.10 Cost Breakdown
4.11 Summary of Tender

Prior to contract between the parties, the set of documents to be agreed on is known as *Tender Documents*. The Consultants are responsible for its drafting. The Quantity Surveyor ("QS") co-ordinates and collates the various parts to form the Tender Documents. There is no fixed form or content in drafting Tender Documents. Depending on the requirements for each project, Tender Documents vary in form and content from project to project.

Essentially, the various parts are drafted to ensure that the scope of work, specifications for material and workmanship, terms, conditions and other requirements are clearly made known to the Contractor.

Upon acceptance of the tender, the set of Tender Documents shall be known as the *Contract Documents* and will be referred to as such in this chapter. Some of the more common parts are discussed as follows:

4.1 Conditions of Contract

The Contract Documents will usually stipulate the *Conditions of Contract* applicable to the parties, e.g. Singapore Institute of Architects Lump Sum Contract, 9th Edition.

These conditions provide for the rights and liabilities of the Owner and Contractor in many common occurring events such as progress payments to the Contractor, adverse weather requiring extension of time, variations in the works required by the Owner or Architect etc.

An example of the public sector conditions of contract, the Public Sector Standard Conditions of Contract for Construction Works, may be viewed at the Building & Construction Authority ("BCA") website.[1]

4.2 Form of Tender

In the Design-Bid-Build ("DBB") system, a *Form of Tender* is a form where a tenderer states the offer of a price which he wishes to be paid to construct and complete the project. An example of a Form of Tender is in Fig. 4.1.

The Form of Tender is an essential document. It states the Tender Sum in clear terms for Owner's acceptance.

4.3 Form for Insurer Bond/Bank Guarantee ("the Guarantee")

Contractors are usually required to provide *the Guarantee* for the amount of 5% of the Contract Sum (i.e. 5% x the lump sum stated in the contract for completion of the Works). The purpose of the Guarantee is to provide security to the Owner in the event of default on the part of the Contractor. It is common for the Guarantee to require the guarantor, i.e. the Bank or Insurance Company, to pay on-demand by the Owner without proof of default on the part of the Contractor.[2]

---

[1] https://www.bca.gov.sg/PSSCOC/psscoc_construction_works.html.
[2] Lim, P. (2015), *Elements of Construction Law in Singapore* (Singapore: Sweet & Maxwell), p. 42.

FORM OF TENDER

To: [Name of Owner]

Having inspected the site, examined the drawings (ref.no.: TD101/D1 to D30)("Drawings"), Conditions of Contract, Specifications, tender documents (ref. no.: TD101)("Tender Documents") and other requirement for the tender for "The Proposed Construction and Completion of a 5-storey building at 101, Dover Road, Singapore", I/We offer to construct, complete and maintain the whole of the Works as shown in Drawings and specified in the Tender Documents for the sum of :-

Singapore Dollars:_____

_____(S$_____).

I/We shall complete the whole of the Works within twelve (12) calendar months from the date of commencement of Works.

The Tender Documents shall consist of the following:

(a) This Form of Tender
(b) Conditions of Contract
(c) [etc...]

The Drawings shall consist of the following:

(a) [etc...]

Name of Tenderer: _____

Contact Person: _____ (Designation :_____)

Tel..no.: _____Fax:_____ Email:_____

Company Stamp: _____ Date: _____

Fig. 4.1   Form of Tender.

Thus, the form and words of the Guarantee are provided in the Tender Documents to ensure that the Contractor complies with obtaining the required on-demand guarantee from an approved Bank or Insurance Company.

4.4 Roofing and other Guarantees or Warranties

Such Guarantees or Warranties are provided by suppliers or proprietary manufacturers or specialist contractors (or Sub-Contractors) to the Owner to safeguard against defects in the products or services.

In particular, it is common for specialist waterproofing contractors to provide Guarantees for a number of years against water leakage arising from defects in the supply and application of their waterproofing compound.

4.5 Preliminaries

Generally, Preliminary items do not form part of the construction work. They are necessary for the proper operation of the work site (see Fig. 4.2). These items are stipulated in the Contract Documents by the Consultants and priced by the Contractor. Some of these items may

Item	Description	*[3]	Amount
1.	The Contractor shall provide, erect and maintain a temporary site office approx. size 37 sq. m to be approved by the Architect. The office should be equipped with all necessary air-conditioning, electricity, water, lighting, telephone service and other services fit for purpose as an office for the duration of the Contract Period.		$20,000.00
2.	The Contractor shall provide and maintain a telephone service to the site for the Contract Period and pay for all charges in connection with the telephone service, including the supply, installation, maintenance and removal.		$3,000.00
3.	The Contractor shall provide, erect and maintain all necessary sheds at the site for the storing of goods, material, equipment and other stores for the purpose of the Works. The design of the sheds shall be approved by the Architect.		$2,000.00
4.	The Contractor shall provide a Performance Bond or Bank Guarantee in the form stipulated in Appendix 1 to the value of 5% of the Contract Sum in favor of the Employer.		$5,000.00

Fig. 4.2   Preliminaries.

---

[3] According to SIA Measurement Contract, 9th Edition, cl 5(2), the Contractor shall indicate "F", "T", "Q" for each of the Preliminary items.

Item	Description	*[3]	Amount
5.	The Contractor shall obtain all guarantees specified in the Tender Documents including the waterproofing guarantees.		Included
	C/F		$30,000.00
	B/F		$30,000.00
6.	The Contractor shall provide and maintain on the Site a signboard including all necessary foundation, words and description to be approved by the Architect. Upon completion of the works, the signboard shall be removed from site.		$5,000.00
7.	The Contractor shall provide and maintain temporary toilets (one for every 20 persons on site) and bathrooms for the duration of the works. The Contractor shall ensure compliance with relevant Authorities and Ministry of Environment laws, regulations and other requirements. If the Contractor fails to do the above, the Contractor shall indemnify the Employer for any fines, losses and damages arising from and in connection with the Contractor's failure.		$10,000.00
8.	The Contractor shall not dump waste materials on any part of the Site. The Contractor shall provide proper bulk bins and containers for waste materials. All rubbish and waste shall be collected and removed from site to Ministry of Environment-approved dumping grounds from time to time.		$10,000.00
9.	The Contractor shall provide and maintain a temporary water supply distribution system for all works on the Site, including Nominated Sub-Contractors' works.		$20,000.00

Fig. 4.2   (*Continued*)

Item	Description	*³	Amount
10.	The Contractor shall provide and maintain temporary lighting and electrical power for all the works including Nominated Sub-Contractors' works.		$100,000.00
	C/F		$175,000.00
	B/F		$175,000.00
11.	The Contractor shall provide and maintain all scaffolding, stagings, planks, walkways, gangways and other necessary platforms for purpose of the works, including Nominated Sub-Contractors' works. Such scaffoldings and other platforms must be approved by the Engineer. The Contractor shall check the installation and sturdiness of the scaffolds and platforms regularly to ensure safety for use.		$50,000.00
12.	The Contractor shall submit the design of, construct and maintain all necessary temporary drains to ensure that rain water and other surface water is properly drained and the site properly drained at all time. The Contractor shall obtain the approval of the Drainage department, Ministry of Environment in respect of the design and construction of such drains.		$10,000.00
13.	The Contractor shall provide and maintain the access road to the site including crossings over drains, channels.		$30,000.00
	[Other items etc...]		[sums priced by Contractor]
	Total carried forward to Item 1, Summary of Tender		$500,000.00

Fig. 4.2   (*Continued*)

be expended by the Contractor prior to the commencement of works, some during the progress of works and some after completion of works. It is necessary to describe in some detail the scope of the Preliminaries so that the Contractor is bound by the requirement specified.

Preliminaries form part of the contractual duties of the Contractor. Hence, it should be clearly and comprehensively drafted. It is common for Preliminary items to comprise many tens of pages in a contract, with each item carefully worded and thoroughly explained to ensure Contractor's understanding of his scope of duty. Disputes may arise if clarity in the contract is not ensured.

Most of the Preliminary items are standard provisions for projects. For the sake of ease of understanding, Fig. 4.2 illustrates some items of Preliminaries, briefly drafted due to space constraint.

The total sum in Preliminaries in Fig. 4.2 is carried forward to Item 1, Summary of Tender (see Fig. 4.11).

4.6 Prime Cost (PC) & Provisional Sum

"PC Sum" stands for *Prime Cost Sum*. When works are referred to be carried out under a "PC Sum", the Owner intends such works to be carried out by Nominated Sub-Contractors or Designated Sub-Contractors (not the Main Contractor).

Where works are to be carried out under a *Provisional Sum*, such works may be carried out by the Main Contractor or Nominated Sub-Contractors or Designated Sub-Contractors. And if such works are not carried out at all, the Provisional Sum should be omitted without being paid to any contractor.

An illustration for PC and Provisional Sum in the Summary of Tender is in Fig. 4.3.

Please refer to Fig. 4.3. During Tendering, the Contractor would state the price for the items 1 to 4 in the Summary of Tender. Items 5 and 6 appear as Provisional Sum and PC Sum respectively.

Item 5 is provided as a Provisional Sum of $50,000 for the supply and installation of kitchen cabinets. The sum of $50,000 is estimated by the Consultant in the Summary of Tender, and hence is not to be

Item	Description	Amount
1.	Allow Preliminaries for the whole project.	[to be priced by Main Contractor]
2.	Carry out Site Preparation.	[as above]
3.	Supply and install Piling to the site.	[as above]
4.	Construct and complete 5-storey building.	[as above]
5.	Allow a Provisional Sum of $50,000.00 for the supply and installation of kitchen cabinets.	$50,000.00
6.	Allow a PC Sum of $100,000.00 for the supply and installation of lifts.	$100,000.00
6(a).	Add: ____% for profit.	[to be priced by Main Contractor]
6(b).	Add: Sum for attendance.	[as above]
	Total carried forward to Form of Tender	

Fig. 4.3   Provisional Sum and PC Sum in Summary of Tender.

priced by the Contractor. Subsequently, the Owner may require the Main Contractor or a Nominated Sub-Contractor or Designated Sub-Contractor to carry out that work or not at all. The actual Sum to be paid may be different from the Provisional Sum estimated, depending on the actual quantity of work carried out and rates assessed in accordance to the contract.

Item 6 is provided as a PC Sum for the supply and installation of lifts. The sum of $100,000 is estimated by the Consultant in the Summary of Tender, and hence is not to be priced by the Contractor. Subsequently, the Owner may require a Nominated Sub-Contractor or Designated Sub-Contractor to carry out that work (not Main Contractor) or not at all. The actual Sum to be paid for carrying out the work by the Nominated Sub-Contractor or Designated Sub-Contractor may be different from the PC Sum, depending on the tender by the Nominated Sub-Contractor or Designated Sub-Contractor for the work. The Main Contractor would be required to state the price for his profit and attendance (item 6(a) and 6(b)) in coordinating and assuming liability over the works of the Nominated or Designated Sub-Contractor.

The Contract Documents should specify "Attendance" from the Main Contractor to the Nominated or Designated Sub-Contractor as in Fig 4.4.

Provisions on the computation of Profit and pricing of Attendance may be stipulated as in Fig 4.5.

1.	The Main Contractor shall provide attendance on the Nominated Sub-Contractors ("NSC") or Designated Sub-Contractors ("DSC") as follows:
(a)	The Main Contractor shall provide all NSC and DSC free and full use of standing scaffolding.
(b)	The Main Contractor shall provide free and full use of canteens, mess rooms and sanitary facilities.
(c)	The Main Contractor shall provide reasonable space to NSC and DSC to erect store-rooms and sheds.
(d)	The Main Contractor shall allow NSC and DSC free use of hoist and elevators.
(e)	The Main Contractor shall provide temporary lighting, electrical power and water for NSC and DSCs' works.
	[Other items etc...]

Fig. 4.4   Main Contractor's attendance to NSC and DSC.

2.	Profit
	The Main Contractor shall insert a percentage in respect of the Profit item under the PC Sum. The amount for profit in respect of a PC Sum item in the Summary of Tender shall be computed as follows:
	Amount of profit = Z% × PC Sum = $Q (where Z is to be fixed by the Main Contractor).
	The Z% shall be used to compute the profit in the Final Account as follows:
	Amount of profit in Final Account = Z% × Adjusted Sub-Contract Sum.
	Attendance
	The amount inserted in attendance for work allowed under a PC Sum shall be a lump sum.

Fig. 4.5   Computation and pricing for Profit and Attendance under PC Sum.

4.7 Particular Specification

Usually, the Architect would specify requirements to the project in the form of *Particular Specification.* There is no fixed method of drafting Particular Specification. A good guide would be to draft the Particular Specification in a "Building Elemental form" or "Building trade form". Hence, the Particular Specification may specify types of materials, standard, installation and other requirements classified according to the Building Elemental form as follows:

(1) Substructure
(2) Superstructure

    (a) RC Frame
    (b) Upper floors
    (c) Roof
    (d) Stairs and ramps
    (e) External Walls
    (f) Windows and external doors
    (g) Internal walls and partitions
    (h) Internal doors

(3) Internal Finishes

    (a) Wall Finishes
    (b) Floor Finishes
    (c) Ceiling Finishes

(4) Fittings Furnishing and Equipment
(5) Services

    (a) Sanitary installation
    (b) Water supply
    (c) Air-conditioning and Mechanical Ventilation
    (d) Electrical Works
    (e) Mechanical Works

4.8 PC Supply Rate

Sometimes, the Owner or Architect has not decided on a particular type of material, e.g. a particular type of floor tiles, at the time of drafting the Tender Documents. Such work may be specified by way of 'PC supply' rate for pricing by the Contractor. When the particular type of material has been selected, the price of material may be adjusted. Please see Fig. 4.6 below as an example.

Item	Description	Unit	Rate	Amount
1.	Supply 300mmx300mm matt homogenous floor tiles (PC Supply rate $25 per m^2).	m^2		$32,000.00
2.	Supply 300mmx100mm high matt homogenous tile skirting (PC Supply rate $5.00 per m).	m		$5,000.00
3.	Labor for laying 300mmx300mm matte homogenous floor tiles laid to pattern on cement and sand screed including pointing with white cement.	m^2		$25,000.00
4.	Ditto but to 100mm high skirting.	m		$9,000.00

Fig. 4.6   Pricing for PC Supply rate.

Items 1 to 4 in Fig. 4.6 provide for Contractor's pricing for the supply and laying of floor tiles. But the specific type of floor tiles has not been specified in Item 1 except that it should cost a PC supply rate of $25 per m^2. Based on the supply rate of $25 per m^2 and quantities measured from drawings, the Contractor prices Item 1 for the supply of floor tiles at $32,000.00 inclusive of wastage, profit and other costs. Similarly, the Contractor prices Item 2 for the supply of skirting tiles at $5,000.00.

The method for adjustment in respect of PC Supply rate may be specified in the Contract Documents as in Fig. 4.7.

When payments are made for Contractor's work in respect of items 1 and 2 of Fig. 4.6, the computation in Fig. 4.8 will apply.

4.9 General Specification

The General Specification may consist of a number of specifications as follows:

- Specification for instrumentation and monitoring with respect to Structural Works;
- Specification for Demolition Works;

Price Adjustment to PC Supply rate

In respect of adjustment to PC Supply rate items, only the amount arising from the PC Supply rate will be omitted. The amount added shall be based on the invoice rate for the supply of material.

Example

Item described in Contract Documents:

"Supply and lay approved 150mm x 150mm x 15mm thick ceramic tiles (PC Supply rate $50 per m² supplied and delivered to site.) bedded in cement and sand mortar (1:3) on screed (measured separately) and pointed in tinted cement."

Say, material cost of ceramic tiles stated in invoice: $60 per m²

Adjustment of rate

(a) Omission

PC Supply rate stated in specification =	$50.00
Add : Wastage 5% =	$ 2.50
	$52.50
Profit & overhead 10%	$ 5.25
Rate for omission	$57.75

(b) Addition

Rate stated in invoice	= $60.00
Add: Wastage 5%	= $ 3.00
	$63.00
Profit & overhead 10%	$ 6.30
Rate for addition	$ 69.30

Fig. 4.7   Adjustment for PC Supply rate specified in Contract Documents.

Payment arising from adjustment to Item 1, Fig. 4.6

(a) Omission

PC Supply rate stated in specification =	$25.00
Add: Wastage 5%	$ 1.25
	$26.25
Profit & overhead 10%	$ 2.63
Rate for Omission	$28.88

(b) Addition

Rate stated in invoice, say =	$40.00
Add: Wastage 5%	$ 2.00
	$42.00
Profit & overhead 10%	$ 4.20
Rate for Addition	$46.20

Hence, payment to Contractor for Item 1 = $32,000 + Q × (46.20 − 28.88), where Q is the quantity of floor area for tiling in m².

Fig. 4.8   Computation for PC Supply rate adjustment.

- Specification for Piling Works;
- Specification for Earthworks;
- Specification for Excavation;
- Specification for Reinforced Concrete Works;
- Specification for Structural Steel Works;
- Specification for Sewerage Works;
- [Etc...]

Soil reports would usually be provided to apprise the Contractor of the soil profile over various part of the site for the purpose of piling works. Usually, it is provided that the Owner does not warrant the accuracy of the soil report and the Contractor is required to conduct his own soil investigation.

There is also a set of National Productivity and Quality Specifications (NPQS) that provides for performance requirements, materials and components, workmanship, verification and submission for various building trades and work.

4.10 Cost Breakdown

You would recall that the Contractor has to submit a Tender Sum in the Form of Tender (see Fig. 4.1). Frequently, the Tender Sum would be a substantial number for the construction and completion of the whole project. Hence, the Contract Documents would require the Tender Sum to be broken down into smaller items and classified under the Elemental form or Building Trade form as follows:

Item	Description	Unit	Rate	Amount
	Element 01: Site Preparation			
1.	Provide PUB-licensed cable detector professional services for locating electrical, water, gas and other existing services at the site and arrange with the relevant authorities for diversion or disconnection of the services.	Sum		$2,000.00
2.	Engage a Professional Engineer ("PE") for the works and submit a report to ensure public safety and prevention of damage to nearby buildings.	Sum		$6,000.00
3.	The Contractor shall provide and maintain temporary surface drainage and silt traps to ensure proper drainage of water to public drains. The surface drainage system shall be designed by a PE and submitted to the relevant authorities for approval.	Sum		-
			C/F	$8,000.00

Fig. 4.9   Cost Breakdown in Building Elemental Form.

Item	Description	Unit	Rate	Amount
			B/F	$8,000.00
4.	The Contractor shall provide a wash bay for vehicles designed by a PE and submitted for approval to the relevant Authorities.	Sum		$5,000.00
	[other items etc...]			
	Any other works not mentioned above but required in the drawings, specifications and other part of the tender documents and law.			[sums priced by Contractor]
	_____ _____ _____			
	Total carried forward to Item 2, Summary of Tender			$50,000.00

Fig. 4.9   (*Continued*)

4.10.1 Building Elemental Form and Pricing in Lump Sum

Usually, the Elemental Form or other format in the Cost Breakdown varies according to the needs of the project. As observed in para 4.1.7, the Building Elemental Form starts with Substructure works. However, in the above, it starts with Site Preparation. Substructure works is reflected later in the Cost Breakdown.

In Fig. 4.9, the Contractor is required to price each item in the column "Amount". Most of the items are priced by the Contractor in the form of a "Sum". In other words, the Contractor has agreed to carry out that item of work for the sum stated. The sum for that item will not be changed due to price increase in inflation or other reasons unless provided in the contract or law. Hence, a Contractor is not entitled to claim a higher sum from the Owner for an item of work because the supplier has raised the price of material.

The total for each Element would be carried forward to the Summary of Tender (see Fig. 4.11).

Item	Description	Unit	Qty	Rate	Amount
	Element 02 — Piling				
1.	Provide a licensed land surveyor to survey, plot and submit plan showing position of piles installed.	Sum			$3,000.00
2.	Install piles under the supervision of the contractor's Professional Engineer and in compliance with Authorities' requirements.	Sum			$4,000.00
3.	Carry out monitoring of settlement and vibration during piling works.	Sum			$3,000.00
4.	Bored Piles				
4a.	Provide bore pile frames, mobilization to and from site and within the site.	Sum			$100,000.00
4b.	Install Bore holes 500mm dia. in ground (In 20 nos.) (Provisional)	m	600	$200	$120,000.00
4c.	Ditto 700mm dia. (In 3 nos.) (Provisional)	m	90	$220	$19,800.00
4d.	Ditto 900mm dia. (In 5 nos.) (Provisional)	m	150	$240	$36,000.00
				C/F	$285,800.00

(*Continued*)

(Continued)

Item	Description	Unit	Qty	Rate	Amount
				B/F	$285,800.00
5a.	Supply and install vibrated reinforced concrete grade 30 to 500mm dia. bored holes. (In 20 nos.) (Provisional)	m	600	$60.00	$36,000.00
5b.	Ditto in 700mm dia. (In 3 nos.) (Provisional)	m	90	$100.00	$9,000.00
5c.	Ditto in 900mm dia. (In 5 nos.) (Provisional)	m	150	$150.00	$22,500.00
6a.	Supply and install cage reinforcement as specified and shown in drawings to 500mm dia. bored holes. (In 20 nos.) (Provisional)	m	300	$25.00	$7,500.00
6b.	Ditto in 700mm dia. (In 3 nos.)(Provisional)	m	45	$$40.00	$1,800.00
6c.	Ditto in 900mm dia. (In 5 nos.) (Provisional)	m	75	$55.00	$4,125.00
7a.	Hack off 500mm dia. bore pile head to expose rebar. Cut and bend rebar all as directed. (Provisional)	No.	20	$150.00	$3,000.00
7b.	Ditto 700mm dia. (Provisional)	No.	3	$170.00	$510.00
7c.	Ditto 900mm dia. (Provisional)	No.	5	$190.00	$950.00
				C/F	$371,185.00

(Continued)

(Continued)

Item	Description	Unit	Qty	Rate	Amount
				B/F	$371,185.00
8a.	Pre-cast Piles Provide piling frames, mobilization to and from site and within the site.	Sum			$90,000.00
8b.	Supply and install 175× 175mm pre-cast concrete piles (In 10 nos.) (Provisional)	m	300	$150.00	$45,000.00
8c.	Ditto for 200×200mm (In 20 nos.) (Provisional)	m	600	$180.00	$108,000.00
8d.	Hack off 175×175 pre-cast concrete pile head to expose rebar. Cut and bend exposed rebar as directed. (In 10 nos.) (Provisional)	No.	10	$200.00	$2,000.00
8e.	Ditto for 200×200mm (In 20 nos.) (Provisional)	No.	20	$250.00	$5,000.00
9.	Provide all necessary plant, equipment, instrument, platform, material and labor for conducting and recording static load test for 1 no. working load test on 900mm dia. bored piles by kentledge method. (Provisional)	Ton	500	$30.00	$15,000.00
				C/F	$636,185.00

(Continued)

(Continued)

Item	Description	Unit	Qty	Rate	Amount
				B/F	$636,185.00
10.	PDA Test	No.	1	$2,500.00	$2,500.00
	[other items etc...]				[sums priced by Contractor]
	Any other works not included in the above but required in the drawings, specifications and other part of the tender documents and law.				
	Total carried forward to Item 3, Summary of Tender				$800,000.00

4.10.2 Provisional Quantity

Please refer to Element 02 — Piling above. Many items in the Piling element have the word "Provisional". To each of these items, "Provisional Quantities" are provided.

For example in Item 4b. in respect of boring 500mm dia. bore holes in the ground, there is an estimated total length of 600m of boring for all 20 bore holes. The estimated 600m of boring is provided by the Consultant in the Cost Breakdown. During preparation of tender, the Contractor had priced $200 per m for the boring of 500mm dia. bore holes computing to the value $120,000 (600mx$200 per m). Subsequently, when the boring is carried out, the actual length of the boring may not be 600m in total. Depending on engineering decision during boring, it may be less, say, 550m in total. Then, the Contractor will be paid 550mx$200 per m for the boring of 500mm dia. bore holes. Hence, the 600m is a provisional quantity as it may be adjusted for actual work done.

The above method is applied to other provisional quantity items, e.g. supply and laying of concrete to the bore holes (items 5a., 5b., 5c.), cage reinforcement bars (items 6a., 6b., 6c.).

Item	Description	Unit	Rate	Amount
	Element 03 - Substructure			
1.	Pile caps including excavation, vibrated reinforced concrete grade 30, formwork and rebar as shown in drawings and as specified.	Sum		$30,000.00
2.	Ditto to Concrete stumps	Sum		$10,000.00
3.	Ditto to Lift pit	Sum		$7,000.00
4.	1st storey (ground) beams including excavation, vibrated reinforced concrete grade 30, formwork and rebar as shown in drawings and as specified.	Sum		$30,000.00
5.	Ditto to 1st storey (ground) slab including lean concrete and hardcore.	Sum		$32,000.00
6.	Supply and install "AH" Elastomeric waterproof membrane to underside of ground slabs, ground beams and lift pit all as specified.	Sum		$12,000.00
7.	Supply and install expansion joints as specified.	Sum		$4,000.00
8.	Supply and install water stops and construction joints as specified.	Sum		included
			C/F	$125,000.00

(Continued)

(Continued)

Item	Description	Unit	Rate	Amount
			B/F	$125,000.00
	[other items etc...]	Sum		[Sums priced by Contractor]
	Any other works not included in the above but required in the drawings, specifications and other part of the tender documents and law.			
	_____			
	_____			
	_____			
	Total carried forward to Collection			$500,000.00

4.10.3 "Collection" Page

As noted in Element 03 — Substructure above, the total sum for the element of Substructure is not carried forward to the Summary of Tender but to a "Collection" page (see Fig. 4.10).

The Substructure and subsequent elements are all part of Item 4 of the Summary of Tender concerning the construction and completion of a 5-storey building. Hence, it may be neater to have an intermediate "Collection" page to sum up the value of the elements and then carry forward to a single Item 4 in the Summary of Tender (see Fig. 4.11).

Generally the Summary of Tender should not be too lengthy. Hence, "Collection" pages are useful in consolidating values of elements or items for transferring to the Summary of Tender.

Item	Description	Unit	Rate	Amount
	<u>Element 04 — Superstructure</u>			
	Supply and install vibrated reinforced concrete grade 30, formwork, rebar to the following as shown in drawings and as specified:	Sum		
1.	1st storey column	Sum		$12,000.00
2.	1st storey RC Wall	Sum		$10,000.00
3.	1st storey Lift shaft wall	Sum		$3,000.00
4.	1st storey staircase	Sum		$5,000.00
5.	2nd storey column	Sum		$11,000.00
6.	2nd RC beam	Sum		$8,000.00
7.	2nd storey RC wall	Sum		$8,000.00
8.	2nd storey lift shaft wall	Sum		$4,000.00
9.	2nd storey staircase	Sum		$4,000.00
10.	3rd storey column	Sum		$12,000.00
11.	3rd storey RC beam	Sum		$7,000.00
12.	3rd storey RC wall	Sum		$7,000.00
13.	3rd storey lift shaft wall	Sum		$3,000.00
			C/F	$94,000.00

(Continued)

(Continued)

Item	Description	Unit	Rate	Amount
			B/F	$94,000.00
14	3rd storey staircase			$3,000.00
	[Other items etc...]			[sums priced by Contractor]
	Any other works not included in the above but required in the drawings, specifications and other part of the tender documents and law. _____ _____ _____			
	Total carried forward to Collection			$2,000,000.00

	Element 05 — Roof and Rainwater goods			
1.	Supply and install proprietary roofing system to RC roof comprising:	Sum		$5,000.00
a)	20mm thick cement and sand screed (1:3) with "AH" waterproofing compound to falls all as specified.	Sum		Included
b)	50mm thick extruded polystyrene insulation foam board as specified.	Sum		Included

(Continued)

(Continued)

Item	Description	Unit	Rate	Amount
c)	One layer of fiber reinforced bituminous felt with 150mm side and end overlaps.	Sum		Included
d)	50mm thick cement and sand (1:3) waterproof with "AH" waterproof compound and finished smooth with steel trowel finish.	Sum		Included
2.	100mm wide scupper drain in panel roofing to falls.	Sum		$500.00
	[Other items etc...]			[Sums priced by contractor]
			C/F	$200,000.00
			B/F	$200,000.00
	Any other works not included in the above but required in the drawings, specifications and other part of the tender documents and law. ————————— ————————— —————————			
	Total carried forward to Collection			$200,000.00

	Element 06 — Wall			
1.	Supply and lay 100mm thick common brick in cement sand mortar bedding (1:3).	Sum		$100,000.00
2.	Ditto 200mm thick.	Sum		$50,000.00
3.	Supply and install 100mm thick lightweight precast panel to wall including framing as specified.	Sum		$30,000.00
4.	Supply and install 10mm thick lightweight gypsum plaster board to wall including framing and 50mm thick fiberglass insulation as specified.	Sum		$10,000.00
5.	[Other items etc...]			
	Any other works not included in the above but required in the drawings, specifications and other part of the tender documents and law.			[Sums priced by Contractor]
	Total carried forward to Collection			$400,000.00

Item	Description	Unit	Rate	Amount
	Element 07 — Door			
	Etc...			
	Total to Collection			$500,000.00
	Element 08 — Windows			
	Etc...			
	Total to Collection			$300,000.00
	Element 09 — Floor Finishes			
	Etc...			
	Total to Collection			$400,000.00
	Element 10 — Wall Finishes			
	Etc...			
	Total to Collection			$300,000.00
	Element 11 — Ceiling Finishes			
	Etc...			
	Total to Collection			$600,000.00
	Element 12 — Sanitary Fittings			
	Etc...			
	Total to Collection			$300,000.00

Collection

Item	Element	Amount
1.	Substructure	$500,000.00
2.	Superstructure	$2,000,000.00
3.	Roof & Rainwater Goods	$200,000.00
4.	Wall	$400,000.00
5.	Door	$500,000.00
6.	Window	$300,000.00
7.	Floor Finishes	$400,000.00
8.	Wall Finishes	$300,000.00
9.	Ceiling Finishes	$600,000.00
10.	Sanitary Fittings	$300,000.00
	Total carried forward to Item 4, Summary of Tender	$5,500,000.00

Fig. 4.10   Collection page.

Summary of Tender

Item	Description	Amount
1.	Allow for Preliminaries for the whole project.	$500,000.00
2.	Carry out Site Preparation.	$50,000.00
3.	Supply and install Piling work to the site.	$800,000.00
4.	Construct and complete the 5-storey building.	$5,500,000.00
5.	Allow a PC Sum of $50,000.00 for the supply and installation of kitchen cabinets.	$50,000.00
5(a).	Add: 1% for profit.	$500.00
5(b).	Add: Sum for attendance.	$1,000.00
6.	Allow a PC Sum of $100,000.00 for the supply and installation of lifts.	$100,000.00
6(a).	Add: 1% for profit.	$1,000.00
6(b).	Add: Sum for attendance.	$1,000.00
7.	Allow a PC Sum of $200,000.00 for supply and installation of electrical works.	$200,000.00
7(a).	Add: 2% for profit.	$4,000.00
7(b).	Add: Sum for attendance.	$2,000.00
	Total carried forward	$7,209,500.00

Fig. 4.11 Summary of Tender.

Fig. 4.11 (*Continued*)

Item	Description	Amount
	Total brought forward	$7,209,500.00
8.	Allow a PC Sum of $300,000.00 for supply and installation of Air-conditioning & Mechanical Ventilation works.	$300,000.00
8(a).	Add: 2% for profit.	$6,000.00
8(b).	Add: Sum for attendance.	$1,000.00
9.	Allow a PC Sum of $100,000.00 for supply and installation of Fire Protection works.	$100,000.00
9(a).	Add: 2% for profit.	$2,000,00
9(b).	Add: Sum for attendance.	$1,000.00
10.	Construct and complete External Works comprising driveway, fencing and landscaping works.	$50,000.00
11.	Allow a Provisional Sum of $800,000.00 for contingency works	$800,000.00
	Total carried forward to Form of Tender	$8,469,500.00

The Summary of Tender provides an overview and consolidation of the Tender Sum. The Tender Sum of $8,469,500.00 would be carried forward to the Form of Tender, the sum representing the price which the Contractor offers to construct and complete the whole of the project in accordance with the Tender Document.

As mentioned earlier, the PC and Provisional Sums are estimated sums allowed in the Summary of Tender. The PC Sums are the estimated value for works to be carried out by NSC and DSC. The Provisional Sum is an estimated value for works which may be carried out by the Main Contractor or NSC. All the other items not estimated in the Summary of Tender must be priced by the Contractor tendering for the works.

The percentage profit for each of the PC Sum shall be inserted by the Contractor in the Summary or Tender. For example in item 9(a), the Contractor inserted "2%" for his profit. Hence, the computation for item 9(a) is $2,000.00 (2% x $100,000.00).

The Contractor should also price a fix sum for attendance in respect to each PC Sum in the Summary of Tender e.g. item 9(b) priced at "$1,000.00" by the Contractor.

The manner of pricing in the Summary of Tender and Cost Breakdown allows a systematic way for computing and assessing Payment Claims. This is discussed later in the book.

CHAPTER 5

Payment Claim

In this chapter, we will discuss the following:

5.1 What is a Payment Claim?
5.2 When to Make a Payment Claim?
5.3 How to Make a Payment Claim?
5.4 Why Make a Payment Claim?

5.1 What is a Payment Claim?

A *Payment Claim* is a claim by the Contractor to the Owner in writing for work done and materials supplied in accordance with the contract. A Payment Claim may also be made by the Sub-Contractor or Supplier to the Main Contractor.[1]

The Contractor's entitlement to Progress Payments, i.e. one-off or regular payments by the Owner to the Contractor for carrying out construction work and supply of materials, is usually provided in the construction contract and is also expressly provided in the Building and Construction Industry Security of Payment Act ("SOP Act").[2] The frequency for Progress Payments is in accordance with the construction contract between Owner and Contractor, and is usually monthly.

In order to facilitate a Progress Payment, the Contractor would first submit a Payment Claim[3] to the Owner. Such a Payment Claim submitted by the Contractor to the Owner provides information as to the amount of payment claimed by the Contractor for work done and materials supplied.

---

[1] SOP Act, s 5.
[2] SOP Act, s 5.
[3] SOP Act, s 10(1).

Then, the Owner (through the assessment of the Consultant QS) would respond with his own verification of work done in respect for that particular Payment Claim and respond with a sum payable to the Contractor.[4]

5.2 When to Make a Payment Claim?

Usually, the construction contract provides for the frequency for making a Payment Claim,[5] which is commonly provided as monthly. Fig. 5.1 shows the time-line for monthly Payment Claims and Payment Responses, i.e. Owner's response to the Contractor on the sum payable upon verifying the Payment Claim. Fig. 5.2 shows the window periods for Payment Claims and Payment Responses.

Contractors would submit their monthly Payment Claims to the Owners during the Contract Period (See Fig. 5.1). Although, there are no restrictions in common standard forms for submitting Payment Claims during the Maintenance Period, it is expected that there would be fewer Payment Claims during the Maintenance Period as compared to during the Contract Period as the works have been completed.

Where a contract does not provide expressly for the time for submission of a Payment Claim then, reg 5(1), SOPR provides,

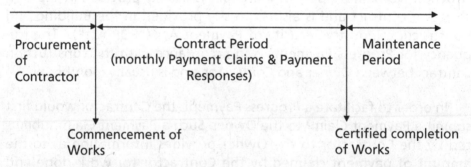

Fig. 5.1   Time-line for monthly Payment Claims and Payment Responses.

---

[4] SOP Act, s 11(1).
[5] For example, SIA Measurement Contract 9th Edition, cl 31(2)(a)(i).

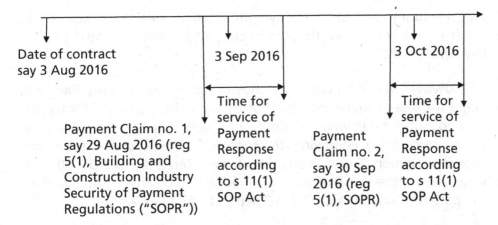

Fig. 5.2 Window periods for Payment Claims and Payment Responses.

"…then a payment claim made under the contract shall be served by the last day of each month following the month in which the contract is made."

In Fig. 5.2, assuming there was no provision in the contract in respect of the time of submission of a Payment Claim, then, according to reg 5(1) SOPR, the submission of the first Payment Claim, if made within the first month of the date of contract, should be by 3 Sep 2016 (being the *last day of each month following the month in which the contract is made,*[6] the contract being made on 3 Aug 2016). In Fig. 5.2, the first Payment Claim was shown to be submitted on 29 Aug 2016 and therefore was properly submitted by 3 Sep 2016.

The second Payment Claim was shown to be submitted on 30 Sep 2016, which was also properly submitted within the second month after the date of contract, the last day of the second month after the date of contract being 3 Oct 2016.

Moving forward, the Payment Claims to be made subsequently on a monthly basis would be the third Payment Claim by 3 Nov, fourth Payment Claim by 3 Dec and so on.

A Contractor may only submit one Payment Claim monthly.[7] Thus, for the monthly period from 3 Aug 2016 to 3 Sep 2016, the Contractor

---

[6] Lee Wee Lick Terence v Chua Say Eng [2012] SGCA 63 at para 94.
[7] Ibid. at para 90.

may not submit two or more Payment Claims. Similarly for each of the other monthly periods, the Contractor may submit at most one (1) Payment Claim.

However, a Contractor may choose not to submit any Payment Claim for one or more months.[8] For example, in Fig. 5.2, a Contractor may not submit a Payment Claim from 3 Aug 2016 to 3 Sep 2016 if he chose not to do so. Following, the contractor, if he chose to do so, may submit a Payment Claim at any time from 3 Sep 2016 to 3 Oct 2016. In which event, the value of work claimed would be from 3 Aug 2016 to the date of claim any time from 3 Sep 2016 to 3 Oct 2016.

5.3 How to Make a Payment Claim?

The information required in a Payment Claim is stipulated in the SOP Act and SOPR.[9] The SOP Act and SOPR do not prescribe any format in reflecting the information required. However, the Building and Construction Authority website provides a guide format.[10]

There is no strict requirement to follow the guide format as long as the SOP Act and SOPR are complied with. Hence, there are many differing formats in the industry.

5.3.1 Covering Letter

A *Covering Letter* is a short and simple letter that accompanies a set of documents and is attached to the documents as the first page. Usually, it highlights and identifies the accompanying documents. An example of a Covering Letter for a Payment Claim is shown in Fig. 5.3. The Covering Letter is sent by the Contractor to the Owner as the front page to the Payment Claim.

The Covering Letter serves as the first page before the Payment Claim. The Covering Letter is followed by the accompanying example of a Payment Claim, as shown in Fig. 5.4.

---

[8] Ibid. at para 90.
[9] SOP Act, s 10(3) and SOPR, reg 5(2).
[10] https://www.bca.gov.sg/SecurityPayment/security_payment_legislation.html as at 8 January 2016.

Contractor's letterhead

Our Ref:
[Date]

[Addressee's name]
[Addressee's address]

*Attention:*_____

Dear Sir

PAYMENT CLAIM NO. 17
PROPOSED ERECTION OF A BUNGALOW HOUSE WITH A BASEMENT
AND A SWIMMING POOL ON LOT 1234 SINGAPORE

We are pleased to submit herewith our Payment Claim no. 17 amounting to
S$_____(Singapore Dollars_____
_____only) excluding GST for the
month ending February 2014.

We look forward to your kind payment as soon as possible. Thank you.

Yours faithfully,

_____
*   [Name]*

Director

Cc: Ace Architects Consultancy — Attention: Mr Andrew Lim
Premium Quantity Surveying & Cost Consultancy — Attention:
Mr David Lim

Fig. 5.3   Covering Letter for Payment Claim.

5.3.2. Payment Claim

A *Payment Claim* is a requirement under the SOP Act. Some of the information required in the Payment Claim form[11] are prescribed by the SOP Act and SOPR.[12] Each Payment Claim is required to state a

---

[11] Guide format at http://www.bca.gov.sg/SecurityPayment/others/SOP_sampleforms.doc.
[12] SOP Act, s 10(3) and SOPR, reg 5(2).

Payment Claim

Payment claim reference number: PC017	Payment claim date: 28/02/2014

To: East West Development Corporation Ltd	
Service address: 101, High Street, Singapore 123456	Tel: 64660088
	Fax: (65) 64668800
	Email: EW@Eastwest.com

Person-in-charge (Respondent): Mr Tan King Kong, Director, Project Management, 81216677 (mobile)

From: Best Contractors Pte Ltd	
Service address: 188, Low Street, Singapore 654321	Tel: 66778899
	Fax: (65) 76548907
	Email: Best@Construct.com

Person-in-charge (Claimant): Joseph Goh, Site Manager, 98767890 (mobile)

Particulars of Contract

Project title: Bungalow at Port Meadow Road	
Contract description: Proposed Construction and Completion of Bungalow House at 8 Port Meadow Road Singapore	Reference period of this claim: From 15/05/2012 to 28/02/2014
Contract number: 045/CONST/2014	
Date of Contract: 2 May 2012	

Fig. 5.4 Payment Claim.

Claim Amount

Value of work done by Main Contractor	$3,479,710.00
Variation	$13,225.00
Value of work done by Nominated Sub-Contractors	$84,500.00
Materials on site	$350,000.00
Less: Retention	($427,743.50)
Total	$3,499,691.50
Less: Previous Payment certified	($3,000,000.00)
CLAIM AMOUNT	$499,691.50

(Claim amount in words)
Singapore Dollars Four hundred and ninety-nine thousand six hundred and ninety-one and cents fifty only

This Payment Claim includes Claim Amount Particulars, breakdown of costs, details, calculation and other supporting documents annexed herein.

No. of pages of this Payment Claim including Claim Amount Particulars, cost breakdown, calculations and supporting document: ###	

Best Contractors Pte Ltd/
Joseph Goh

Name of claimant/authorized representative: ――――――――――

Authorized signature &   XXXXXXXXXXXXXX

Organization stamp (if applicable):   XXX

――――――――――

Date: 28/02/2014

――――――――――

Fig. 5.4   (*Continued*)

Claim Amount based on the valuation of work done that is supported by calculations and attachments.[13]

The particulars of the Claim Amount are on the next page.

5.3.3 Claim Amount Particulars

Please refer to figure on *Claim Amount Particulars*. In the Claim Amount Particulars, the "Description" provides a short description of the work to be claimed. Usually, the "Description" column lists the type of work itemized in the Summary of Tender.

The values under "Contract Value" are values priced by the Contractor for the respective items in the Summary of Tender.

"% Work Done" shows the percentage of work done for each of the item described.

The values under "Claim" are obtained from computations in the detail Sections mentioned under "Remarks".

The total value of work done and materials delivered to site was $3,927,435.00. The Retention Sums are then deducted (based on cls 31(7) and 31(8)[14]). If the computed Retention Sums do not exceed 5% of the Contract Sum, the computed Retention Sums are deducted as shown in the figure. If the computed Retention Sums exceed 5% of the Contract Sum, the Owner may only deduct the maximum of 5% of the Contract Sum.[15]

GST[16] should not be added into the Payment Claim. Subsequently, the Contractor may include GST when the invoice is issued.

The Previous Payment certified is deducted to arrive at the Claim Amount of $499,691.50.

---

[13] SOP Act, s 10(3).
[14] SIA Measurement Contract, 9th Edition.
[15] SIA Measurement Contract, 9th Edition, cl 31(8).
[16] website: GST: Construction Industry-IRAS. Para 2.2.

Claim Amount Particulars

Item	Description	Contract value	% Work done	Claim	Remarks
1.0	Preliminaries	$631,000.00	67.83%	$428,020.00	See Section 1.0
2.0	Schedule of works				
2.1	Piling	$133,400.00	100%	$137,090.00	See Section 2.1
2.2	Site works	$124,600.00	100%	$124,600.00	See Section 2.2
2.3	Bungalow house	$10,070,000.00	27.71%	$2,790,000.00	See Collection
3.0	NSC for Air-conditioning and Mechanical Works	$134,000.00	48.89%	$65,500.00	See Section 3.0
4.0	NSC for Electrical Works	$78,000.00	24.36%	$19,000.00	See Section 4.0
5.0	Variations	NA		$13,225.00	See Section 5.0
	Total			$3,577,435.00	
	Value of work done by Main Contractor (Items 1.0, 2.1, 2.2, 2.3, 5.0)			$3,492,935.00	
	Value of work done by NSC (Items 3.0, 4.0)			$84,500.00	
	Materials on site			$350,000.00	See invoice attached
	Total			$3,927,435.00	
	Less: Retention for value of work done (10% x $3,577,435.00)		($357,743.50)		
	Less: Retention for material on site (20% x $350,000.00)		($70,000.00)	($427,743.50)	
	Total			$3,499,691.50	
	Less: Previous payment certified			($3,000,000.00)	
	Claim Amount			**$499,691.50**	

As mentioned earlier, there is no mandatory format in reflecting the Claim Amount Particulars and computation. The format shown in the foregoing figure is one of many. The objective in showing the computation is to satisfy the requirement of the SOPR.[17]

Further, the SOPR requires a breakdown of the Claim Amount. Hence, the "Remark" column provides a detailed breakdown of the Claim Amount in the Sections. The detailed breakdown for the claim for Preliminaries may be found in Section 1.0 hereafter.

Section 1.0   Preliminaries

Item	Description	Contract value	% Completed	Claim
1.	Insurance	$10,000.00	100%	$10,000.00
2.	Performance Bond	$10,000.00	100%	$10,000.00
3.	Site Safety	$9,000.00	66%	$5,940.00
4.	Setting out of works	$14,000.00	100%	$14,000.00
5.	Restrict nuisance	$7,500.00	66%	$4,950.00
6.	Vector control	$9,000.00	66%	$5,940.00
7.	Temp access and temp roads	$10.500.00	66%	$6,930.00
8.	Hoarding, fencing and gates	$8,000.00	66%	$5,280.00
9.	Project name board	$3,000.00	66%	$1,980.00
10.	Scaffolding	$50,000.00	66%	$33,000.00
11.	Contractor's temp office, stores	$20,000.00	66%	$13,200.00
12.	Temp water and electricity	$50,000.00	66%	$33,000.00
13.	Pant, tools, vehicles	$250,000.00	66%	$165,000.00
14.	Removal of rubbish	$20,000.00	66%	$13,200.00
15.	Testing of materials	$10,000.00	66%	$6,600.00
16.	Labor	$150,000.00	66%	$99,000.00
	Total Claim carried forward to Claim Amount Particulars	$631,000.00		$428,020.00

---

[17] Building and Construction Industry Security of Payment Regulation (Rg1, GN No S2/2005, r 5(2).

5.3.4 Preliminaries

Please refer to Section 1.0 *Preliminaries*. Again, the Author wishes to reiterate that the format and information provided in the figure are not mandatory as long as there is compliance with the SOP Act and SOPR. There are many formats with different other information being reflected.

In the Author's view, the payment for Preliminaries should be based on the value of Preliminaries expended through the construction period. The value of Preliminaries expended with time is usually not a straight line formula.

For the sake of simplicity in the figure on Preliminaries, the Author has adopted two-thirds (66%) expenditure for Preliminaries by assuming a straight line expenditure formula for the duration of the Contract Period.

In reality, some Preliminary items would have been expended by the Contractor before commencement of works, e.g. Payment of premiums for insurance. Hence, it may be in order to claim and be paid for such items. Other items have an initial capital outlay and maintenance costs during the contract period and payment for Preliminaries may be made on that basis too.

You would also note that the total claim for Preliminaries of $428,020.00 in the above figure is carried forward to Item 1 of the Claim Amount Particulars.

5.3.5 Piling

5.3.5.1 Lump Sum

Please refer to Section 2.1 Piling (Items A to K). The information under "Description" "Unit", and "Rate" in the figures would have been obtained from the cost breakdown or Summary of Tender in the Contract Documents. You would notice the word "Sum" in the "Unit" column. In other Unit columns, there are "m" and "No." representing "meter" and "number" respectively. Where the word "Sum" was provided, the Contractor would be required to price a lump sum for that item of work described. For example, for items A and C, the Contractor priced them at $55,000.00 and $13,000.00 respectively in the cost

Section 2.1   Piling

Item	Description	Unit	Qty	Rate	Contract value	Qty completed	Claim
	Piling Equipment						
A	Allow for supply and mobilization	Sum	—	—	$55,000.00	100%	$55,000.00
B	Allow for demobilization and removal	Sum	—	—	Included		Included
	Instruments and monitoring						
C	Provide, protect and monitor with piling instruments for period specified	Sum			$13,000.00	100%	$13,000.00
D	Provide, protect and monitor with noise measuring instruments	Sum			Included		Included
						C/F	$68,000.00

	Particulars					B/F	
	Provide and install Pre-cast concrete jack-in piles (Provisional)						$68,000.00
E	175x175mm piles (30 nos.)(Provisional)	m	250	$23.00	$5,750.00	230	$5,290.00
F	200x200mm piles (150 nos.)(Provisional)	m	1,500	$32.00	$48,000.00	1,650	$52,800.00
	Cutting heads of piles						
G	175x175mm piles (Provisional)	No.	30	$30.00	$900.00	20	$600.00
H	200x200mm piles (Provisional)	No.	150	$35.00	$5,250.00	140	$4,900.00
	Static Load Test						
I	200x200mm piles (Provisional)	No.	1	$2,500.00	$2,500.00	1	$2,500.00
	Dynamic Analysis Test						
J	175x175mm piles (Provisional)	No.	1	$1,500.00	$1,500.00	1	$1,500.00
K	200x200mm piles (Provisional)	No.	1	$1,500.00	$1,500.00	1	$1,500.00
	Total carried forward to Payment Claim						$137,090.00

breakdown or Summary of Tender. And the Contractor has completed fully those two items of work and has claimed 100%.

5.3.5.2 Provisional quantities

In Items E to K, Section 2.1, the "Unit" column shows "m" or "No.". The "Qty" column shows numerical numbers for meter or numbers. These numerical numbers are "Provisional Quantities" estimated by the Consultant and fixed for each item in the Summary of Tender. Naturally, the Provisional Quantity for an item of work may not be the actual quantity of work done at the end of the day. Particularly in Piling work, engineering decisions are made during the work as to the sufficiency of the length of the piles driven into the ground. It may end up shorter or longer than the Provisional Quantities.

Hence, in Item E, the total completed pile length was 230m, shorter than the Provisional Quantity of 250m. The Contractor was only entitled to claim $5,290 for that item (230m x $23per m). In Item F, the total completed pile length was 1,650m, which was longer than the Provisional Quantity of 1,500m. The Contractor was entitled to claim $52,800.00 for that item (1,650m x $32 per m).

5.3.6 Site Works

Please refer to Section 2.2 on Site Works. As explained earlier, the information under "Description", "Unit", "Contract Values" would have been obtained from the cost breakdown or Summary of Tender. In the figures, the Contractor has claimed 100% for work done for each item.

5.3.7 Bungalow House

Please refer to the figures in Section 2.3 Bungalow House. The Concrete Work items consisting of lean concrete, reinforced concrete, formwork, rebar and waterproof membrane were lump sum items from the cost breakdown or Summary of Tender. The Contractor, having completed them, claimed 100% payment for each of the item.

Similarly for Brickwork and Blockwork, Roofing, Carpentry and Joinery works, these were completed and the Contractor claimed 100% payment except for Item B Timber Door, which the Contractor claimed 50%.

Section 2.2   Site Works

Item	Description	Unit	Contract value	% of work done	Claim
	Demolition				
A	Demolish and remove off-site existing building, structures and facilities	Sum	$8,800.00	100%	$8,800.00
	Excavation				
B	Clear the site, including cutting down trees and removing them off-site	Sum	$3,400.00	100%	$3,400.00
C	Excavate, fill and compact as specified and remove excess earth off-site	Sum	$29,000.00	100%	$29,000.00
D	Backfill and compact with approved earth as specified	Sum	$55,000.00	100%	$55,000.00
	Water disposal				
E	Keep the site, including the excavation, free of water by pumping or other methods	Sum	$2,000.00	100%	$2,000.00
				C/F	$98,200.00
				B/F	$98,200.00
	Termite Treatment				
F	Provide anti-termite soil treatment to surface in contact with building structure	Sum	$6,400.00	100%	$6,400.00
	Boundary wall and fencing				
G	Provide 1800mm high chain-link fence	Sum	$20,000.00	100%	$20,000.00
	Total carried forward to Claim Amount Particulars				$124,600.00

Section 2.3   Bungalow House

Item	Description	Unit	Contract value	% of work done	Total work done
	Concrete work				
	Lean concrete				
A	Basement and ground slabs	Sum	$150,000.00	100%	$150,000.00
B	Pile-caps	Sum	$20,000.00	100%	$20,000.00
	Reinforced concrete grade 30				
C	Pile-caps	Sum	$30,000.00	100%	$30,000.00
D	Basement/ground slabs	Sum	$150,000.00	100%	$150,000.00
E	Suspended slabs	Sum	$100,000.00	100%	$100,000.00
F	Retaining walls	Sum	$300,500.00	100%	$300,500.00
G	Columns	Sum	$50,500.00	100%	$50,500.00
H	Beams	Sum	$70,500.00	100%	$70,500.00
I	Staircases and landings	Sum	$20,500.00	100%	$20,500.00
	Formwork				
J	Pile-caps	Sum	$60,500.00	100%	$60,500.00
K	Basement/ground slabs	Sum	$130,000.00	100%	$130,000.00
L	Suspended slabs	Sum	$130,500.00	100%	$130,500.00
M	Retaining walls	Sum	$50,500.00	100%	$50,500.00
N	Columns	Sum	$30,200.00	100%	$30,200.00
O	Beams	Sum	$50,200.00	100%	$50,200.00
P	Staircases and landings	Sum	$10,500.00	100%	$10,500.00

(*Continued*)

Section 2.3 (*Continued*)

Item	Description	Unit	Contract value	% of work done	Total work done
	Reinforcement bars				
Q	Pile-caps	Sum	$80,300.00	100%	$80,300.00
R	Basement/ground slabs	Sum	$300,000.00	100%	$300,000.00
S	Suspended slabs	Sum	$120,000.00	100%	$120,000.00
T	Retaining walls	Sum	$90,000.00	100%	$90,000.00
U	Columns	Sum	$70,000.00	100%	$70,000.00
V	Beams	Sum	$120,000.00	100%	$120,000.00
W	Staircases and landings	Sum	$50,500.00	100%	$50,500.00
	Waterproof membrane				
X	Waterproof admixture to concrete	Sum	$250,000.00	100%	$250,000.00
Y	Water-stops in concrete structures	Sum	$150,500.00	100%	$150,500.00
	Total Claim carried forward to Collection				$2,367,000.00

(*Continued*)

Section 2.3   (*Continued*)

Item	Description	Unit	Contract value	% of work done	Amount Claimed
	<u>Brickwork and Blockwork</u>				
A	100mm thick	Sum	$96,000.00	100%	$96,000.00
B	150mm thick	Sum	$130,000.00	100%	$130,000.00
	Total Claim forward to Collection				$226,000.00
A	<u>Roofing</u> Aluminum roof	Sum	$35,000.00	100%	$35,000.00
B	Rainwater downpipes, gutter	Sum	$73,000.00	100%	$73,000.00
	Total Claim carried forward to Collection				$108,000.00
	<u>Carpentry and Joinery</u> <u>Timber cladding</u>				
A	25mm thick balau cladding to column c/w frames	Sum	$41,500.00	100%	$41,500.00
B	<u>Timber Doors</u> Single leaf timber door c/w frames	Sum	$95,000.00	50%	$47,500.00
	Total Claim carried forward to Collection				$89,000.00

5.3.7.1 Collection page

Each of total claim for Concrete Work, Brickwork and Blockwork, Roofing, Carpentry and Joinery is carried forward to a "Collection" page for consolidation (see next figure). The Collection page serves as an intermediate consolidation page so that the total of various items may be added together and the aggregate number carried forward to the Payment Claim Particulars as illustrated below.

Collection

Item	Description	Value
1.	Concrete Work	$2,367,000.00
2.	Brickwork and blockwork	$226,000.00
3.	Roofing	$108,000.00
4.	Carpentry and Joinery	$89,000.00
	Total carried forward to Claim Amount Particulars	$2,790,000.00

Section 3.0   NSC for Air-conditioning and Mechanical Works

Item	Description	Unit	Contract value	% work done	Amount Claimed
A	Supply and install 1 lot of VRV air-conditioners system	Sum	$55,000.00	50%	$27,500.00
B	Supply and install 1 lot supply and return aircon ductwork	Sum	$50,000.00	50%	$25,000.00
C	Supply and install 1 lot aircon pipes	Sum	$24,000.00	50%	$12,000.00
D	Supply and install 1 lot drainage pipes	Sum	$2,000.00	50%	$1,000.00
E	Provide twelve (12) months maintenance service	Sum	$2,400.00	0	0
F	Test and commission aircon system	Sum	$1,000.00	0	0
	Total carried forward to Payment Claim Particulars		$134,000.00		$65,500.00

Section 4.0   NSC for Electrical Works

Item	Description	Unit	Contract value	% work done	Amount Claimed
A	Supply and install 1 lot lighting system	Sum	$20,000.00	0	0
B	Supply and install 1 lot garden lighting system	Sum	$3,000.00	50%	$1,500.00
C	Supply and install 1 lot power outlet distribution system	Sum	$15,000.00	50%	$7,500.00
D	Supply and install 1 lot ceiling fan point	Sum	$2,000.00	0	0
E	Supply and install 1 lot main switchboard and 1 lot distribution board	Sum	$20,000.00	50%	$10,000.00
F	Supply and install 1 lot lightning protection system	Sum	$10,000.00	0	0
G	Supply and install 1 lot SCV system	Sum	$1,000.00	0	0
H	Supply and install 1 lot MIO TV system	Sum	$2,000.00	0	0
I	Engage LEW to prepare submission and turn-on	Sum	$3,000.00	0	0
J	Test and commission	Sum	$2,000.00	0	0
	Total carried forward to Claim Amount Particulars		$78,000.00		$19,000.00

5.3.8 NSC for Air-conditioning and Mechanical Works and NSC for Electrical Works

Please refer to the above two figures in Section 3.0 NSC for Air-conditioning and Mechanical Works and Section 4.0 NSC for Electrical

Works. Contractually, NSCs' works are deemed part of the Main Contractor's work in the Main Contract between Owner and Main Contractor. The Main Contractor is liable to the Owner for the work done by NSCs. The Main Contractor is also entitled to payment from the Owner for work done by NSCs. After being paid, the Main Contractor would pay the respective NSCs for their works.

Thus, the Main Contractor, in the Claim Amount Particulars, has included a claim for NSC work for Air-conditioning and Mechanical Works and NSC for Electrical Works in Items 3.0 and 4.0 respectively.

The breakdown for the NSCs' claims is found in Section 3.0 and 4.0 above. The Item Descriptions and Contract values are obtained from the cost breakdown or Summary of Tender of the NSC Sub-contracts. The Sub-Contractors will claim the "% work done" and submit their claims to the Main Contractor. The Main Contractor will consolidate the NSCs claims together with his own claim and detailed breakdown. The consolidated claim will be submitted to the Owner as a Payment Claim as in Fig. 5.4.

Section 5.0   Particulars of Variations

Item	Description	Instruction ref.	Value of variation claim	Remarks
1.	Changes to Pile Cap	AI-1	$5,600.00	See
2.	Changes to Letter Box	AI-2	$125.00	measurement
3.	Changes to Driveway	AI-3	$2,500.00	sheets and
4.	Changes to timber flooring	AI-4	$3,500.00	rates attached.
5.	Changes to pump room	AI-5	$2,500.00	
6.	Changes to Main Entrance	AI-6	($1,000.00)	
7.	Add Shower door	AI-7	(deleted)	
	Total carried forward to Claim Amount Particulars		$13,225.00	

5.3.9 Variation

The above figure in Section 5.0 shows the breakdown for the Variation claim. The Total Claim of $13,225.00 is being carried forward to the Payment Claim Particulars.

Variations are changes to the contracted works during the construction period. For example, in Item 1 Pile Cap, there was an Instruction reference AI-1 requiring the Contractor to change the construction of the Pile Cap according to new design. Arising from the change, the Contractor has claimed $5,600.00 additional payment.

Variation may also result in omission, as may be seen in Item 6 Change to Main Entrance.

5.3.10 Remarks on how to make a Payment Claim

Although there is no mandatory requirement in the format of a Payment Claim, convention dictates that it has an overall summary of the Claim Amount. The items making up the Claim Amount would be broken down into smaller items of works in separate sheets and further broken down if necessary to satisfy the requirement of the SOP Act and SOPR.

Each of the broken down items would be claimed by the Contractor by assessing the % work done and computed to obtain the Claim Amount.

Hence, the format of the Payment Claim, including the Claim Amount Particulars and detailed breakdown shows a systematic and easy to understand assessment and computation of the Claim Amount. This is important in the overall scheme of the SOP Act as it allows the Owner (through the QS) to verify the assessment and computation of the Payment Claim and to arrive at the Owner's response (with QS's assistance) to the Claim Amount. The Payment Response will be discussed in the next chapter.

5.4 Why Make a Payment Claim?

A Payment Claim is a statutory term under SOP Act. The SOP Act was enacted to assist Contractors, Sub-Contractors, Suppliers and others described in the Act to obtain:

> ""*a fast and low cost adjudication system to resolve payment disputes*"".[18]

Hence, many aggrieved contractors have applied for adjudication for recovery of payments. However, it was established in *Terence v Chua Say Eng*,[19] a Court of Appeal decision, that without a valid Payment Claim, there could be no valid Adjudication Determination under the SOP Act as the purported Adjudication Determination would be set aside for lack of jurisdiction of the Adjudicator.

Thus, to avail themselves to the recovery process in Adjudication, the Contractors and Suppliers must submit a Payment Claim, in all aspects satisfying the requirements of the SOP Act.

---

[18] see *Singapore Parliamentary Debates, Official Report* (16 November 2004) vol 78 at col 1112 (Cedric Foo Chee Keng, Minister of State for National Development).
[19] [2012] SGCA 63.

CHAPTER 6

Payment Response and Payment Certificate

In this chapter, we will discuss the following:

6.1 What is a Payment Response?
6.2 When to Make a Payment Response
6.3 How to Make a Payment Response
6.4 Why Make a Payment Response?
6.5 Practice of Making Payment Claims and Payment Responses in the Construction Industry
6.6 What is a Payment Certificate?
6.7 When to Issue a Payment Certificate
6.8 How to Issue a Payment Certificate

6.1 What is a Payment Response?

A Payment Response is a response from the Owner to the Contractor's Payment Claim.[1] You recall in Chapter 5 that the Contractor would submit Payment Claims claiming amounts for his work done and materials on site. On receipt of each Payment Claim, the Owner must make his own assessment of the amount he would pay. This assessment is known as the Payment Response. Although responded in the name of the Owner to the Contractor, the Payment Response is actually assessed and computed by the QS.

6.2 When to Make a Payment Response

Please refer to Figs. 5.1 and 5.2. For each Payment Claim submitted, the SOP Act requires the Owner to respond with a Payment Response.

---

[1]SOP Act, s 11(1).

The time prescribed[2] by the SOP Act to serve a Payment Response is as follows:

(a) By the date specified or determined in the Contract or within 21 days after the Payment Claim is served, whichever is earlier.
(b) Where the Contract does not contain a provision for the time of service of the Payment Response, within 7 days after the Payment Claim is served.

In default of a time provision in the Contract for the service of Payment Response (as mentioned in (b) above), the SOP Act prescribes that the Payment Response must be served within 7 days after the Payment Claim is served. This 7-day period is a very short time for the preparation of a Payment Response. Time would be required to visit the site to assess the value of work done and materials on site, followed by subsequent computation and drafting of the valuation report. Further, the SOP Act's definition of the word 'day' means any day other than a public holiday within the meaning of the Holidays Act. That means the 7-day period includes Saturdays and Sundays, as Saturdays and Sundays are not public holidays under the Holidays Act. Thus, the QS preparing the Payment Response must practically complete the site visit, assessment, computation, report and service of the Payment Response within 5 days of receipt of the Payment Claim (as Saturdays and Sundays are non-working days).

Hence, most construction contracts would expressly provide for the maximum 21-day period for the service of the Payment Response (as mentioned in (a) above) to maximize the time period for the service of the Payment Response.

Apart from the s 11(1) SOP Act prescribing the time of service of the Payment Response to within 21 days after the Payment Claim is served, the s 12(2) SOP Act allows the Payment Response to be filed by the end of the Dispute Settlement Period.[3]

---

[2]SOP Act, s 11(1)(a), (b).
[3]W Y Steel Construction Pte Ltd v Osko Pte Ltd [2013] 3 SLR 380; [2013] SGCA 32, para 24.

6.3 How to Make a Payment Response

The information required in a Payment Response is stipulated in the SOP Act and SOPR.[4] There is no prescribed format for a Payment Response in the SOP Act or SOPR. But the BCA website provides a guide format.[5] An example of a Payment Response in response to the Payment Claim in Chapter 5 is given in Fig. 6.1.

Payment Response

Payment Response reference number: PR017	Payment Response date: 19/03/2014

To: Best Contractors Pte. Ltd.	
Service address: 188, Low Street, Singapore 654321	Tel: 66778899
	Fax: (65) 76548907
	Email: Best@Construct.com

Person-in-charge (Claimant): Joseph Goh, Site Manager, 98767890 (mobile)

From: East West Development Corporation Ltd.	
Service address: 101, High Street, Singapore 123456	Tel: 64660088
	Fax: (65) 64668800
	Email: EW@Eastwest.com

Person-in-charge (Respondent): Mr. Tan King Kong, Director, Project Management, 81216677 (mobile)

Particulars of Contract

Project title: Bungalow at Port Meadow Road	
Contract description: Proposed Construction and Completion of Bungalow House at 8 Port Meadow Road Singapore Contract number: 045/CONST/2014 Date of Contract: 2 May 2012	Reference period of this claim: From 15/05/2012 to 28/02/2014

Fig. 6.1(a)   Payment response.

---

[4]SOP Act, s 11(3). SOPR, reg 6(1).
[5]www.bca.gov.sg/SecurityPayment/security_payment_legislation.html

Fig. 6.1(a)　*(Continued)*

Payment Claim Identification

Payment Claim reference no.: PC017

Payment Claim date: 28/02/2014

Claim Amount: $499,691.50

Response Amount

Value of work done by Main Contractor	$3,201,000.00
Variation	$10,000.00
Value of work done by Nominated sub-contractors	$70,000.00
Materials on site	$350,000.00
Less: Retention	($398,100.00)
Total	$3,232,900.00
Less: Previous Payment certified	($3,000,000.00)
RESPONSE AMOUNT	$232,900.00

(Amount in words)

Singapore Dollars Two hundred and thirty-two thousand, and nine hundred only.

This Payment Response includes the Response Amount Particulars, breakdown of costs, details, calculation and other supporting documents annexed herein.

No. of pages of this Payment Response including Response Amount Particulars, cost breakdown, calculations and supporting document: ### pages

Name of respondent/authorized representative:	East West Development Corporation Ltd/Tan King Kong
Authorized signature and	XXXXXXXXXXXXXX
Organization stamp (if applicable):	XXX
Date:	*19/03/2014*

6.3.1 Payment Response

A Payment Response is a requirement under the SOP Act in response to a Payment Claim.[6] It states an amount for which the Employer would pay in respect of a particular Payment Claim made by the Claimant.[7] Usually, the verification of the Payment Claim and assessment of the Response Amount are computed by the QS. The information required in the Payment Response is prescribed by the SOP Act and SOPR.[8] Each Payment Response should state the Response Amount as shown in the above Payment Response.

Where the Response Amount is less than the Claimed Amount, there should be further breakdown and details of the Response Amount as follows[9]:

(a) The amount that the Respondent proposes to pay for each item of the Claimed Amount;
(b) The reasons for the difference;
(c) The calculations which show how the amount that the Respondent proposes to pay is derived;
(d) The amount that is being withheld, the reasons for doing so, and the calculations which show how the amount being withheld is derived.

See Response Amount Particulars in Fig. 6.1(b).

6.3.2 Goods and Services Tax ("GST")

Under the Goods and Services Tax Act, the Contractor is required to charge GST in Progress Payments as follows[10]:

(a) Based on the payment received or invoice issued, whichever is earlier;
(b) Based on the gross amount of the Contract Sum.

Based on the IRAS GST: Construction Industry guide, GST should not be computed in the Payment Claim or Payment Response. GST should be included when the actual payment is paid to the Contractor and in the invoice (mentioned in (a) above).

---

[6]SOP Act, s 11(1).
[7]SOP Act, s 11(3)(b).
[8]SOP Act, s 11(3). SOPR, reg 6(1).
[9]SOPR, reg 6(1)(d).
[10]Website: GST:Construction_Industry -IRAS, para 2.2.

Response Amount Particulars

Item	Description	Contract Value	Claim Amount	Response Amount	Remarks
1.0	Preliminaries	$631,000.00	$428,020.00	$400,000.00	See Section 1.0
2.0	Schedule of Works				
2.1	Piling	$133,400.00	$137,090.00	$137,090.00	See Section 2.1
2.2	Site works	$124,600.00	$124,600.00	$124,600.00	See Section 2.2
2.3	Bungalow house	$10,070,000.00	$2,790,000.00	$2,539,310.00	See Collection
3.0	NSC for Air-conditioning and Mechanical Works	$134,000.00	$65,500.00	$55,000.00	See Section 3.0
4.0	NSC for Electrical Works	$78,000.00	$19,000.00	$15,000.00	See Section 4.0
5.0	Variations	NA	$13,225.00	$10,000.00	See Section 5.0
	Total		$3,577,435.00	$3,281,000.00	

Value of work done by Main Contractor (Items 1.0, 2.1, 2.2, 2.3, 5.0)	$3,492,935.00	$3,211,000.00
Value of work done by NSC (Items 3.0, 4.0)	$84,500.00	$70,000.00
Materials on site	$350,000.00	$350,000.00 — See invoice attached
Total	$3,927,435.00	3,631,000.00
Less: Retention for value of work done (10% x $3,281,000)	($357,743.50)	($328,100.00) [$3,211,000 +$70,000 =$3,281,000]
Less: Retention for material on site (20% x $350,000.00)	($70,000.00)	($70,000.00)
Total	$3,499,691.50	$3,232,900.00
Less: Previous payment certified	($3,000,000.00)	($3,000,000.00)
Final Amount	$499,691.50	$232,900.00

Fig. 6.1(b)   Response amount particulars.

6.3.3 Response Amount Particulars

Please refer to figure on Response Amount Particulars above. You would notice that the format follows the Payment Claim submitted by the Claimant in Fig 5.4. To each item of the Payment Claim, the Respondent provides his assessment and value in the Payment Response.

For example, in respect of the Claimant's claim in Section 1.0 Preliminaries (Fig. 6.1(c)) at $428,020.00, the Respondent has valued the same item at $400,000.00 in the Response Amount. There is a reduction of $28,020.00. By way of comparing the values claimed and values responded, we are able to identify the very items where the differences lie, i.e. items 13 and 16. In Note 1/13 and Note 1/16, the reasons for the differences and calculations are given for the derivation of the respective response amount.

There are further differences in items 2.3, 3.0, 4.0 and 5.0. Based on the detailed breakdown for Section 2.3, 3.0, 4.0 and 5.0, we would be able to identify the items that give rise to the differences in values, the reasons for differences and corresponding calculations (albeit in measurement sheets not shown).

The calculation of other parts of the Response Amount Particulars should correspond to the items in the Claim Amount Particulars (Fig 5.4). The Response Amount of $232,900.00 corresponds to the Claimed Amount of $499,691.50.

6.3.3.1 Payment to Nominated Sub-Contractors

In addition to the Response Amount payable to the Main Contractor, it would also be advisable to derive the amount payable by the Main Contractor to the Nominated Sub-Contractors. Based on valuation or work done and computation in the Response Amount Particulars in the above figure, Nominated Sub-Contractors' value of work done is also included in the payment to the Main Contractor.

The payment allocated to the Nominated Sub-Contractor for Air-conditioning and Mechanical Works may be computed as follows:

Value of work done	$55,000.00
Less: Retention (10%)	$ 5,500.00
	$49,500.00
Less: Previous certification	$30,000.00
Payment to NSC for Air-conditioning and Mechanical Works	$19,500.00

The payment allocated to Nominated Sub-Contractor for Electrical Works may be computed as follows:

Value of work done	$15,000.00
Less: Retention (10%)	$ 1,500.00
	$13,500.00
Less: Previous certification	$10,000.00
Payment to NSC for Electrical Works	$ 3,500.00

Hence, the Respondent should also make known to the Claimant (Main Contractor) and the Nominated Sub-Contractors so that they are aware of the amounts payable to the Nominated Sub-Contractors from the Response Amounts.

6.3.4 Piling

Please refer to Section 2.1 Piling. Although there are no differences in the Claim and Response Amounts for Piling, it may be in good order, for completeness sake, to show the Claim Amounts and Response Amounts for each item.

6.3.5 Site Works

Please refer to Section 2.2 Site Works. Although there are no differences in the Claim and Response Amounts for Site Works, it may be in good order, for completeness sake, to show the Claim Amounts and Response Amounts for each item.

Section 1.0   Preliminaries

Item	Description	Contract value	Claim Amount	Response Amount	Reasons for differences
1.	Insurance	$10,000.00	$10,000.00	$10,000.00	
2.	Performance Bond	$10,000.00	$10,000.00	$10,000.00	
3.	Site Safety	$9,000.00	$5,940.00	$5,940.00	
4.	Setting out of works	$14,000.00	$14,000.00	$14,000.00	
5.	Restriction of nuisance	$7,500.00	$4,950.00	$4,950.00	
6.	Vector control	$9,000.00	$5,940.00	$5,940.00	
7.	Temporary access and roads	$10,500.00	$6,930.00	$6,930.00	
8.	Hoarding, fencing and gates	$8,000.00	$5,280.00	$5,280.00	
9.	Project name board	$3,000.00	$1,980.00	$1,980.00	
10.	Scaffolding	$50,000.00	$33,000.00	$33,000.00	

11.	Contractor's temp office, stores	$20,000.00	$13,200.00	
12.	Temporary water and electricity	$50,000.00	$33,000.00	
13.	Plant, tools, vehicles	$250,000.00	$145,000.00	Note 1/13
14.	Removal of rubbish	$20,000.00	$13,200.00	
15.	Testing of materials	$10,000.00	$6,600.00	
16.	Labor	$150,000.00	$90,980.00	Note 1/16
	Total Amount carried forward to Response Amount Particulars	$631,000.00	$428,020.00	

Fig. 6.1(c)   Response amount to preliminaries.

Reasons for differences in Section 1.0 Preliminaries

Note	Reasons for differences
1/13	Based on the Contractor's schedule dated XX/XX/2014 for plants, tools and vehicles on site, the current percentage of the value of plant, tools and vehicles employed on site is only 58% and not 66% of the item priced in the Preliminaries. (i.e. 58% x $250,000.00 = $145,000.00)
1/16	Based on the Contractor's schedule dated XX/XX/2014 for current labor force, the labor employed on site is only 60.6% and not 66% (i.e. 60.65% x $150,000.00 = $90,980.00)

Fig. 6.1(d)   Reasons for differences in preliminaries.

Section 2.1   Piling

Item	Description	Unit	Qty.	Rate	Contract value	Claim amount	Response amount
	Piling Equipment						
A	Allow for supply and mobilization	Sum	—	—	$55,000.00	$55,000.00	$55,000.00
B	Allow for demobilization and removal	Sum	—	—	Included	Included	—
	Instruments and monitoring						
C	Provide, protect and monitor with piling instruments for period specified	Sum			$13,000.00	$13,000.00	$13,000.00
D	Provide protect and monitor with noise measuring instruments	Sum			Included	Included	—
	Total carried forward					$68,000.00	$68,000.00

	Total brought forward					$68,000.00	$68,000.00
E	Provide and install Pre-cast concrete jack-in piles (Provisional)						
	175 x 175mm piles (30 nos.) (Provisional)	m	250	$23.00	$5,750.00	$5,290.00	$5,290.00
F	200 x 200mm piles (150 nos.) (Provisional)	m	1,500	$32.00	$48,000.00	$52,800.00	$52,800.00
	Cutting heads of piles						
G	175 x 175mm piles (Provisional)	No.	30	$30.00	$900.00	$600.00	$600.00
H	200 x 200mm piles (Provisional)	No.	150	$35.00	$5,250.00	$4,900.00	$4,900.00

(Continued)

Section 2.1   (*Continued*)

Item	Description	Unit	Qty.	Rate	Contract value	Claim Amount	Response Amount
	Static Load Test						
I	200 x 200mm piles (Provisional)	No.	1	$2,500.00	$2,500.00	$2,500.00	$2,500.00
	Dynamic Analysis Test						
J	175 x 175mm piles (Provisional)	No.	1	$1,500.00	$1,500.00	$1,500.00	$1,500.00
K	200 x 200mm piles (Provisional)	No.	1	$1,500.00	$1,500.00	$1,500.00	$1,500.00
	Total carried forward to Response Amount Particulars					$137,090.00	$137,090.00

Fig. 6.1(e)   Response amount to piling.

Section 2.2   Site Works

Item	Description	Unit	Contract value	Claim Amount	Response Amount
	Demolition				
A	Demolish and remove off-site existing building, structures and facilities	Sum	$8,800.00	$8,800.00	$8,800.00
	Excavation				
B	Clear the site including cutting down trees and removing them off-site	Sum	$3,400.00	$3,400.00	$3,400.00
C	Excavate, fill and compact as specified and remove excess earth off-site	Sum	$29,000.00	$29,000.00	$29,000.00
D	Backfill and compact with approved earth as specified	Sum	$55,000.00	$55,000.00	$55,000.00

(Continued)

Section 2.2   (Continued)

Item	Description	Unit	Contract value	Claim amount	Response amount
	Water disposal				
E	Keep the site including the excavation free of water by pumping or other methods	Sum	$2,000.00	$2,000.00	$2,000.00
	Total carried forward			$98,200.00	$98,200.00
	Total brought forward			$98,200.00	$98,200.00
	Termite Treatment				
F	Provide anti-termite soil treatment to surface in contact with building structure	Sum	$6,400.00	$6,400.00	$6,400.00
	Boundary wall and fencing				
G	Provide 1800mm high chain-link fence	Sum	$20,000.00	$20,000.00	$20,000.00
	Total carried forward to Claim Amount Particulars			$124,600.00	$124,600.00

Fig. 6.1(f)   Response amount to site works.

Section 2.3   Bungalow House

Item	Description	Unit	Contract value	Claim Amount	Response Amount	Reason for differences
	Concrete work					
	Lean concrete					
A	Basement and ground slabs	Sum	$150,000.00	$150,000.00	$150,000.00	
B	Pile-caps	Sum	$20,000.00	$20,000.00	$20,000.00	
	Reinforced concrete grade 30					
C	Pile-caps	Sum	$30,000.00	$30,000.00	$30,000.00	
D	Basement/ground slabs	Sum	$150,000.00	$150,000.00	$150,000.00	
E	Suspended slabs	Sum	$100,000.00	$100,000.00	$100,000.00	

(Continued)

Section 2.3   *(Continued)*

Item	Description	Unit	Contract value	Claim Amount	Response Amount	Reason for differences
F	Retaining walls	Sum	$300,500.00	$300,500.00	$250,000.00	Note 2.3/F
G	Columns	Sum	$50,500.00	$50,500.00	$50,500.00	
H	Beams	Sum	$70,500.00	$70,500.00	$70,500.00	
I	Staircases and landings	Sum	$20,500.00	$20,500.00	$20,500.00	
	Formwork					
J	Pile-caps	Sum	$60,500.00	$60,500.00	$60,500.00	
K	Basement/ground slabs	Sum	$130,500.00	$130,000.00	$130,000.00	
L	Suspended slabs	Sum	$130,500.00	$130,500.00	$130,500.00	
M	Retaining walls	Sum	$50,500.00	$50,500.00	$40,000.00	Note 2.3/M
N	Columns	Sum	$30,200.00	$30,200.00	$30,200.00	
O	Beams	Sum	$50,200.00	$50,200.00	$50,200.00	
P	Staircases and landings	Sum	$10,500.00	$10,500.00	$10,500.00	

	Reinforcement bars					
Q	Pile-caps	Sum	$80,300.00	$80,300.00	$80,300.00	
R	Basement/ground slabs	Sum	$300,000.00	$300,000.00	$300,000.00	
S	Suspended slabs	Sum	$120,000.00	$120,000.00	$120,000.00	
T	Retaining walls	Sum	$90,000.00	$90,000.00	$63,310.00	Note 2.3/T
U	Columns	Sum	$70,000.00	$70,000.00	$70,000.00	
V	Beams	Sum	$120,000.00	$120,000.00	$120,000.00	
W	Staircases and landings	Sum	$50,500.00	$50,500.00	$50,500.00	
	Waterproof membrane					
X	Waterproof admixture to concrete	Sum	$31,300.00	$31,300.00	$31,300.00	
Y	Water-stops in concrete structures	Sum	$150,500.00	$150,500.00	$150,500.00	
	Total Claim carried forward to Collection			$2,367,000.00	$2,279,310.00	

Fig. 6.1(g)   Response amount to bungalow house.

Reasons for differences in Section 2.3 Bungalow House

Note	Reasons for differences
2.3/F	The retaining wall is not completed. The RC work is about 83% completed. See measurement sheet 1 attached.
2.3/M	The formwork to retaining wall is not completed. See measurement in sheet 2 attached.
2.3/T	The rebar to retaining wall is not completed. See measurement sheet 3 attached.

Fig. 6.1(h)   Reasons for differences in bungalow house.

Section 2.3   Bungalow House   (Continued)

Item	Description	Unit	Contract value	Claimed Amount	Response Amount	Reasons for differences
	Brickwork and Blockwork					
A	100mm thick	Sum	$96,000.00	$96,000.00	$40,000.00	Note 2.3/Brickwork/A
B	150mm thick	Sum	$130,000.00	$130,000.00	$130,000.00	
	Total Claim forward to Collection			$226,000.00	$170,000.00	
	Roofing					
A	Aluminum roof	Sum	$35,000.00	$35,000.00	$30,000.00	Note 2.3/Roofing/A
B	Rainwater downpipes, gutter	Sum	$73,000.00	$73,000.00	$20,000.00	Note 2.3/Roofing/B
	Total Claim carried forward to Collection			$108,000.00	$50,000.00	
	Carpentry and Joinery					
A	Timber cladding 25mm thick balau cladding to column c/w frames	Sum	$41,500.00	$41,500.00	$20,000.00	Note 2.3/Carpentry/A
B	Timber doors Single leaf timber door c/w frames	Sum	$95,000.00	$47,500.00	$20,000.00	Note 2.3/Carpentry/B
	Total Claim carried forward to Collection			$89,000.00	$40,000.00	

Fig. 6.1(i)   Response amount to bungalow house.

Reasons for differences in Section 2.3 Bungalow House

Note	Reasons for differences
2.3/Brickwork/A	Poor workmanship to be rectified by removal and relaying of brickwork. See measurement sheet 4.
2.3/Roofing/A	Flashing to eaves not installed. See measurement sheet 5.
2.3/Roofing/B	Rainwater down pipe and gutter not installed. See measurement sheet 6.
2.3/Carpentry/A	Balau cladding not installed. See measurement sheet 7.
2.3/Carpentry/B	Timber doors installed were defective with cracks. 20% of the doors were accepted for payment as shown in drawing no. Arch/door/4.

Fig. 6.1(j) Reasons for differences in bungalow house.

6.3.6 Bungalow House

Please refer to the figures in Section 2.3 Bungalow House above.

We notice the differences in Claim Amounts and Response Amounts in:

(a) Concrete work items F, M and T;
(b) Brickwork item A;
(c) Roofing items A and B;
(d) Carpentry items A and B.

As required in the SOPR, the reasons for differences and calculations should be given as provided above.

6.3.7 Collection Page

For completeness sake, Response Amount should be provided in the Collection page as shown below. The reasons for the differences had already been given earlier.

Collection

Item	Description	Claimed Amount	Response Amount
1.	Concrete Work	$2,367,000.00	$2,279,310.00
2.	Brickwork and blockwork	$226,000.00	$170,000.00
3.	Roofing	$108,000.00	$50,000.00
4.	Carpentry and Joinery	$89,000.00	$40,000.00
	Total carried forward to Response Amount Particulars	$2,790,000.00	$2,539,310.00

Fig. 6.1(k)   Collection page for response amount to bungalow house.

Section 3.0   Nominated Sub-Contract for Air-conditioning and Mechanical Works

Item	Description	Unit	Contract value	Claimed Amount	Response Amount	Reasons for differences
A	Supply and install 1 lot of VRV air-conditioners system	Sum	$55,000.00	$27,500.00	$27,500.00	
B	Supply and install 1 lot supply and return air-con ductwork	Sum	$50,000.00	$25,000.00	$25,000.00	
C	Supply and install 1 lot air-con pipes	Sum	$24,000.00	$12,000.00	$2,500.00	Note 3/C
D	Supply and install 1 lot drainage pipes	Sum	$2,000.00	$1,000.00	0	Note 3/D
E	Provide twelve (12) months maintenance service	Sum	$2,400.00	0	0	
F	Test and commission air-con system	Sum	$1,000.00	0	0	
	Total carried forward to Response Amount Particulars		$134,400.00	$65,500.00	$55,000.00	

Fig. 6.1(I)   Response amount to NSC for air-conditioning and mechanical works.

Reasons for differences in Section 3.0 Nominated Sub-Contract for Air-conditioning and Mechanical Works

Note	Reasons for differences
3/C	Air-con pipes not installed. Only 10.4% installed. (10.4% x $24,000)
3/D	Drainage pipes not installed at all.

Fig. 6.1(m) Reasons for differences in response amount to NSC for air-conditioning and mechanical works.

Section 4.0 Nominated Sub-Contract for Electrical Works

Item	Description	Unit	Contract value	Claimed amount	Response amount	Reasons for differences
A	Supply and install 1 lot lighting system	Sum	$20,000.00	0	0	
B	Supply and install 1 lot garden lighting system	Sum	$3,000.00	$1,500.00	$1,500.00	
C	Supply and install 1 lot power outlet distribution system	Sum	$15,000.00	$7,500.00	$5,500.00	Note 4/C
D	Supply and install 1 lot ceiling fan point	Sum	$2,000.00	0	0	
E	Supply and install 1 lot main switch-board and 1 lot distribution board	Sum	$20,000.00	$10,000.00	$8,000.00	Note 4/E
F	Supply and install 1 lot lightning protection system	Sum	$10,000.00	0	0	
G	Supply and install 1 lot SCV system	Sum	$1,000.00	0	0	
H	Supply and install 1 lot MIO TV system	Sum	$2,000.00	0	0	
I	Engage LEW to prepare submission and turn-on	Sum	$3,000.00	0	0	
J	Test and commission	Sum	$2,000.00	0	0	
	Total carried forward to Response Amount Particulars		$78,000.00	$19,000.00	$15,000.00	

Fig. 6.1(n)   Response amount to NSC for electrical works.

Reasons for differences in Section 4.0 Nominated Sub-Contract for Electrical Works

Note	Reasons for differences
4/C	About 1/3 of the power outlet distribution system was installed. (0.36 x $15,000)
4/E	About 40% of the main switchboard and distribution board was installed. (40% x $20,000)

Fig. 6.1(o)   Reasons for differences in response amount to NSC for electrical works.

Section 5.0   Particulars of Variations

Item	Description	Instruction ref.	Claim Amount	Response Amount	Reasons for differences
1.	Changes to pile cap	AI-1	$5,600.00	$5,600.00	
2.	Changes to letter box	AI-2	$125.00	0	Note 5/2
3.	Changes to driveway	AI-3	$2,500.00	0	Note 5/3
4.	Changes to timber flooring	AI-4	$3,500.00	$2,900.00	Note 5/4
5.	Changes to pump room	AI-5	$2,500.00	$2,500.00	
6.	Changes to main entrance	AI-6	($1,000.00)	($1,000.00)	
7.	Add shower door	AI-7	(deleted)	—	
	Total carried forward to Response Amount Particulars		$13,225.00	$10,000.00	

Fig. 6.1(p)   Response amount to variation works.

	Reasons for the differences for Response Amount to Variation Works
Note	
5/2	Change to letter box not done.
5/3	Change to driveway was defective due to failure to compact hardcore in accordance with instruction AI-3. Cracks appeared due to settlement. Reconstruct driveway.
5/4	Not complete with polishing to surface as required in the AI-4. Deduct 200m^2 x $3.00 = $600.00.

Fig. 6.1(q)   Reasons for differences in response amount to variation works.

6.3.8 Nominated Sub-Contract for Air-conditioning and Mechanical Works and Nominated Sub-Contract for Electrical Works

Please refer to the above two figures on Section 3.0 Nominated Sub-Contract for Air-conditioning and Mechanical Works and Section 4.0 Nominated Sub-Contract for Electrical Works. Response Amounts and reasons for differences are as shown in the above figures.

6.3.9 Variation

The above figure in Section 5.0 shows the Claim Amount and corresponding Response Amount and reasons for differences.

6.3.10 Remarks on how to make a Payment Response

In compliance with the SOP Act and SOPR, the format of the Payment Response should provide the Response Amount to each item of the Claimed Amount. In this way, differences are easily spotted and the reasons for the difference and calculations for the derivation of the Response Amount are clearly made known.

6.4 Why Make a Payment Response

A Payment Response is a requirement under SOP Act in response to a Payment Claim. In providing a Payment Response, one is able to identify the differences between the Claimed Amount and Response Amount and the corresponding reasons and calculations for the difference.

In the event of adjudication, the Adjudicator would be able to understand clearly the differences and reasons and reach a determination in resolving the dispute.

6.5 Practice of Making Payment Claims and Payment Responses in the Construction Industry

Where there is a substantial difference between the Claim Amount and Response Amount, the Contractor will not be satisfied as he will be paid substantially below his claim. If dissatisfied with the Payment Response, the Contractor may make an Adjudication Application in accordance with the SOP Act. Naturally, Owner and Consultants are reluctant to go through the process of adjudication. In order to avoid adjudication, the QS may process the Payment Claim as follows:

6.5.1 Joint valuation of Payment Claim between QS and Contractor

Prior to a Payment Claim, both the QS and the Contractor would together carry out a site valuation, and agree on the value of work done, materials on site and the Claim Amount. Then the Contractor would submit the Payment Claim according to the values and Claim Amount agreed. As the values had been agreed on with the QS, there should not be any difference in the Response Amount compared to the Claim Amount. The QS would assess a Payment Response with the same values as the Payment Claim.

6.5.2 Valuation of Payment Response between QS and Contractor

Where the Contractor submits a Payment Claim without prior agreement on the Claim Amount, the QS would subsequently carry out a site valuation of work done and materials on site so as to derive the Response Amount on his own accord. Then, prior to the issue of the Payment Response, the QS would have a meeting with Contractor to explain the differences between the Claim Amount and Response Amount. With the Contractor's understanding in the valuation and derivation of the Response Amount, it is expected that the Contractor would react less adversely to the Response Amount, and hopefully, avoid an Adjudication Application.

The above two methods are known to be applied in the construction industry to avoid substantial differences in Payment Claim and Response.

6.6 What is a Payment Certificate?

A Payment Certificate (also known as 'Interim Certificate', 'Progress Payment Certificate') is a certificate by the Architect or other authorised persons under the contract certifying the amount due and payable to the Contractor.[11] Except for any deduction allowed under the contract, the Owner should upon receipt of the Payment Certificate pay the amount due to the Contractor.[12]

What is the difference between a Payment Certificate and Payment Response? Both appear to reflect the payment due and payable by the Owner to the Contractor. More often than not, the Amount payable certified in a Payment Certificate is the same as the Response Amount. For that reason, PSSCOC provides that the Payment Certificate shall be deemed the Payment Response if the Owner does not provide a separate response in compliance with the SOP Act.[13] Where a standard form contract does not provide accordingly, there are usually expressed provisions in the Contract Documents to state that the Payment Certificate is deemed a Payment Response in accordance with the SOP Act. Naturally, such a Payment Certificate must satisfy all requirements prescribed in the SOP Act and SOPR.

Historically, the Payment Certificate certified by the Architect (or other authorised persons) is accepted for the purpose of making progress payments. In preparing a Payment Certificate, the QS carries out a valuation of work done and materials delivered to site. Thereafter, the QS forwards the valuation, computation and amount payable to the Architect. Then, the Architect certifies the Payment Certificate accordingly and issues it to the Contractor.

In a Payment Response, more is required than just valuing and computing the amount payable to the Contractor. The Owner (with QS's assessment in compliance with the SOP Act) is required to give

---

[11] SIA Measurement Contract, 9th Edition, cl 31(3). PSSCOC 2014 Edition, cl 32.2(1).
[12] Tropicon Contractors Pte Ltd v Lojan Properties Pte Ltd [1989] 1 SLR 610; [1989] 3 MLJ 216.
[13] PSSCOC 2014 Edition, cl 32.2(2).

reasons and show the calculation in deriving an amount if the amount which he proposes to pay is less than the amount claimed.

Thus, in the SIA Measurement Contract 9th Edition, cl 31(3) requires the Architect to issue an Interim Certificate within 14 days of receipt of the Payment Claim. Then, he should issue a Payment Response within 21 days after service of the Payment Claim in accordance with cl 31(15)(a). In all likelihood, both documents will reflect the same amount payable to the Contractor, but the Payment Response will show more details as required under the SOP Act. Therefore, it is not surprising that expressed provisions are contained in the Contract Documents to state that the Payment Certificate will be deemed the Payment Response to reduce duplication in both documents.

6.7 When to Issue a Payment Certificate

Both the SIA Measurement Contract 9th Edition[14] and the PSSCOC 2014 Edition[15] provide for issue of Payment Certificates within 14 days of receipt of Payment Claim from the Contractor. As the Payment Claims are usually submitted monthly, the Payment Certificates should be certified and issued monthly to the Contractor.

6.8 How to Issue a Payment Certificate

Upon receipt of a Payment Claim from the Contractor, the QS would carry out a site visit to value the work done and materials on site. Then the QS would carry out the assessment and computation and forward his valuation including computation of the amount payable to the Architect. Upon receipt of the valuation and amount payable from the QS, the Architect would certify payment accordingly.

An example of a Payment Certificate is in Fig. 6.2.

---

[14] cl 31(3).
[15] cl 32.2(1).

Corresponding to each Payment Certificate issued by the Architect to the Main Contractor, the Main Contractor may also be required to issue Sub-Contract Payment Certificates to the Nominated Sub-Contractors. The valuation in the Sub-Contract Payment Certificates would be based on the valuation certified to the Main Contractor.

In Fig. 6.1(b) Response Amount Particulars, we note the following:

<u>Response Amount ($)</u>

Item 3.0	NSC for Air-conditioning and Mechanical Works	55,000.00
Item 4.0	NSC for Electrical Works	15,000.00

Based on the Response Amount, the QS would compute (in para 6.3.3.1) the amount payable by the Main Contractor to the respective Nominated Sub-Contractors:

- Payment to NSC for Air-Conditioning and Mechanical Works $19,500.00
- Payment to NSC for Electrical Works $3,500.00

The QS would make known to both the Main Contractor and the respective Nominated Sub-Contractors in respect of the above valuations and payments. Thus, based on the QS valuation and payment, the Main Contractor would certify Sub-Contract Payment Certificates to the respective Nominated Sub-Contractors.

INTERIM CERTIFICATE NO. 10
13 March 2014

PROPOSED BUNGALOW HOUSE AT 8 FIR TREE ROAD, SINGAPORE
123456

Contractor: Best Contractor Pte. Ltd.
Address: XXX

Pursuant to Clause 31 (3) of the Contract Conditions, I hereby certify
payment to the Contractor as follows:

Value of work done by Contractor	$10,554,345.22
Value of materials on site	$ —
Value of work done by Nominated Sub-Contractors	$2,500,000.00
Less: Retention Sum (10%)	($1,305,434.52)
	$11,748,910.70
Less: Sums previously certified	($10,500,234.40)
Amount due (excluding GST)	$1,248,676.30

Certified by:

_____

Name:
Architect

Distribution to:

Fig. 6.2   Payment certificate.

CHAPTER 7

Variations

In this chapter, we will discuss the following:

7.1 What is a Variation?

A Variation is a change in the original contract works, materials or methods of working.

When a project is tendered on a DBB delivery method, the works and materials are fixed by the drawing and specifications in the tender documents. Tenderers are required to tender a price to construct and complete according to the tender documents.

However, during the contract period, Owners and Consultants may wish to change parts of the project. The Architect may do so by issuing a Variation Instruction (sometimes called a "Variation Order") to the Contractor. The Contractor must carry out the Variation Instruction or order. Generally, such a Variation Instruction may give rise to additional payment (sometimes a reduction in payment) to the Contractor and in some circumstances may also result in Extension of Time.

7.2 Instructions and Directions in the SIA Form of Contract

There are two (2) types of orders that the Architect may give to the Contractor under the SIA form of contract, namely:

(a) Instruction;
(b) Direction

Both the Instruction and Direction[1] must be carried out by the Contractor. In issuing an Instruction, there is in principle, additional payment as well as an increase or reduction in the Contract Sum. In issuing a Direction, there is neither additional payment nor increase in the Contract Sum, but in some cases it may result in a reduction of the Contract Sum.

Generally, a Direction is issued to direct the Contractor to remedy a default by him. Hence, no additional payment would be paid to the Contractor in compliance with the Direction. An example is issuing a Direction to demolish and reconstruct a defect caused by the poor workmanship of the Contractor.

On the other hand, an Instruction is issued to require the Contractor to change certain works as desired by the Architect or Owner, not due to any default on the part of the Contractor. This type of Instruction is known as a Variation Instruction. Hence, such changes come with additional payment, Extension of Time or reduction in payment (due to reduction in the value of work) to the Contractor. An example is issuing an Instruction to change the original painting finish to ceramic wall tiles for ornamental or other reasons (not due to a default of the Contractor). Naturally, additional payment (and even Extension of Time) may be expected in this Instruction.

Examples of an Architect's Direction and Instruction are given in Fig. 7.1 and Fig. 7.2 respectively.

---

[1]SIA Measurement Contract, 9th Edition, cl 1.

ARCHITECT'S DIRECTION
[Clause 1(1) Conditions of Building Contract]

Issued date: 6 May 2015

To: Best Contractor Pte. Ltd.
No. 8 Kaki Bukit Avenue 6
Singapore 654321

Attention: Mr. Peter Tan

Project: PROPOSED ERECTION OF TERRACE HOUSES AT 8 MAPLE ROAD SINGAPORE

Project No.: LE(D) 924

I refer to the Letter of Award dated 1 April 2015 and Conditions of Contract. You have failed to provide the following documents:

(1) Construction program for the project;
(2) Copies of insurance policy;
(3) Bank guarantee for the sum of $1,500,000.00 from an approved bank or insurance company.

You are hereby directed to submit the above documents immediately for our approval.

Issued by:

_____(Sgd)_____
[Name   ]
Architect

Cc: [Employer]
[QS]

_____

We acknowledge receipt of the above-mentioned Direction dated 6 May 2015.

_____(Sgd)_____
Name: [   ]
Designation: [   ]
Contractor's stamp: [   ]
Date: [   ]

Fig. 7.1   Architect's Direction.

ARCHITECT'S INSTRUCTION
[Clause 1(1) Conditions of Building Contract]

Issued date: 10 October 2015

To: Best Contractor Pte. Ltd.
No. 8 Kaki Bukit Avenue 6
Singapore 654321

Attention: Mr. Peter Tan

**Project: PROPOSED ERECTION OF TERRACE HOUSES AT 8 MAPLE
ROAD SINGAPORE**

Project No.: LE(D) 924

You are hereby instructed with immediate effect as follows:

(1) Change the painting finish in all toilet walls to ceramic wall tiles in
accordance with the drawings attached. You shall submit samples
of ceramic wall tiles for the Architect's approval prior to execution
of works.

(2) Increase the length of the precast drains to connect to the public
drain as shown in drawing attached.

Issued by:

_____(Sgd)_____
[Name   ]
Architect

Cc: [Employer]
[QS]

_____

We acknowledge receipt of the above-mentioned Instruction dated
10 October 2015.

_____(Sgd)_____
Name: [   ]
Designation: [   ]
Contractor's stamp: [   ]
Date: [   ]

Fig. 7.2   Architect's Instruction.

7.3 Instructions in the PSSCOC Form of Contract

Under the PSSCOC Form of Contract, the Superintending Officer ("SO") gives an order to the Contractor by an *Instruction*.[2] The Instruction in the PSSCOC Form of Contract does not have the same character as an Instruction in the SIA Form of Contract. When the SO issues an Instruction under the PSSCOC Form of Contract, there may or may not be additional payment in carrying out the Instruction, i.e. it may or may not be a Variation Instruction. Hence, it is necessary to confirm in writing with the SO that the Instruction given was a Variation Instruction entitling the Contractor to additional payment.[3]

In the PSSCOC Form of Contract, there is no "Direction" as defined in the SIA Form of Contract.

7.4 Valuation of Variations

The valuation of Variations is based on measurement of quantities of work and application of the relevant rates to the quantities of work. For example:

"Change the painting finish in all toilet walls to ceramic wall tiles in accordance with the drawings attached."

The first step is to value the omission of painting work from the Contract Sum as the Contractor will no longer be carrying out this work. The quantity (in meter square) of painting to be omitted should be measured from the drawing in accordance with the Code of Practice for Construction Electronic Measurement Standards (CEMS) as required under the SIA Form of Contract.[4]

If the original painting work as specified consists of primer, undercoat and finishing coat, the rates ($ per meter square) for primer, undercoat and finishing coat are applied to the quantity of painting work omitted. If such rates are found in the Contract Documents

---

[2] PSSCOC Form of Contract, 2014 Edition, cl 2.5.
[3] PSSCOC Form of Contract, 2014 Edition, cl 19.2.
[4] SIA Measurement Contract, 9th Edition, cl 13(1).

(priced by Contractor in the Bills of Quantities or available in the Fixed Schedule of Rates), then such rates would be applicable. Where the rates are not available in the Contract Documents or for some reasons not applicable,[5] then there are usually provisions in the Conditions of Contract to derive the relevant rates.

Hence, the valuation of omission of painting works is obtained by applying the relevant rates to the quantities of painting work to be omitted.

The next step is the valuation of the additional ceramic wall tile work. The quantity for wall tiling work is measured from the drawing. Then, the relevant rate for supplying and installing of ceramic wall tile is applied to the quantity if the rate is available in the Contract Documents. If the rate is not available in the Contract Documents, then the rate of the ceramic wall tile work will be derived based on other provisions in the Contract.

The valuation for this Variation is as follows:

Valuation for Variation in changing wall painting to wall tile =
+ valuation for the addition of ceramic wall tile works
− valuation for the omission of wall painting works

The above valuation would be used to adjust the Contract Sum in determining the adjusted Contract Sum in light of this variation.

7.5 Rates for Variations

The Conditions of Contract usually provide for a system[6] in obtaining rates for Variations.

In Measurement Contracts (where there are Bills of Quantities), the BQ rates would be applied to derive the valuation of Variation

---

[5]SIA Measurement Contract, 9th Edition, cl 12(4). PSSCOC 2014 Edition, cl 20.1.
[6]SIA Form of Contract, Measurement Contract, 9th Edition, cl 12(4). PSSCOC Form of Contract, 2014 Edition, cl 20.1.

where the Variation work items have been priced by the Contractor in the BQ.

In Lump Sum Contracts, the Contract Documents usually contain a Schedule of Rates or Fixed Schedule of Rates for many common items in the building trade (see Fig. 7.5 and 7.6 FSR). These rates are to be applied in the valuation of Variation. Tenderers entering into a Lump Sum Contract are required to agree to these rates for purpose of valuation of Variation.

Where there are no equivalent items in the BQ or Schedule of Rates in respect of a particular Variation item, the Conditions of Contract usually provide that, wherever possible, the rates in the Contract Documents be used as a basis for deriving the rates for Variation work items.[7]

Where it is not possible to use the rates in the BQ or Schedule of Rates as a basis for deriving the rates for Variation work items, the Conditions of Contract usually provide that the QS apply a *fair market rate*[8] for the Variation work item.

A common method of obtaining a fair market rate is to obtain quotations from independent sources in respect of a particular work item. Usually, one would get three independent quotations and compute the mean value as the fair market rate.

Where fair market valuation is not possible, then the valuation may be obtained by building up the cost components for the item.[9] This method of computing the rate is called the *Star Rate*. The Star Rate is obtained by building up *actual cost* components of a work

---

[7]SIA Form of Contract, Measurement Contract, 9th Edition, cl 12(4)(b),(c). PSSCOC Form of Contract, 2014 Edition, cl 20.1(b).
[8]SIA Form of Contract, Measurement Contract, 9th Edition, cl 12(4)(d). PSSCOC Form of Contract, 2014 Edition, cl 20(1)(c).
[9]SIA Form of Contract, Measurement Contract, 9th Edition, cl 12(4)(e)(ii). PSSCOC Form of Contract, 2014 Edition, cl 20.1(d).

item. For example, the Star Rate of a made-to-order cabinet shown in drawing is as follows:

• Actual material cost paid or to be paid	:$XX
• Actual labor cost paid or to be paid	:$XX
• Actual plant & equipment cost paid or to be paid	:$XX
Sub-total	$XX
Add: Overheads and profit	
(15% of Sub-total)	$XX
Star Rate	$XX

Fig. 7.3   Star Rate computation.

As could be seen in Fig. 7.3 Star Rate computation, the Star Rate is a build-up of component rates such as cost of material, cost of labor, cost of plant and equipment. A factor of 15% is added to the aggregate as overheads and profit. Naturally, the components in the Star Rate should be only those mentioned in the provisions.[10] The objective of a Star Rate is to build up a realistic rate for a work item based on the cost of basic components.

A basic difficulty in computing a Star Rate is the requirement that the component costs must be costs *actually incurred (paid or to be paid)*. Then, the Contract provides for the 15% margin over costs actually incurred.

7.6 Sample Valuation of Variations

There are usually many Variations in a project. It is common to have a summary of all the Variation orders and particulars of each Variation order as shown in Figs 7.4, 7.5 and 7.6.

The summary of Variation Orders ("VO") in Fig. 7.4 consists of all the VO, i.e. VO 1 to VO 12. It gives the description of each VO, the Contractor's claim, QS Assessment, Agreed Amount and the Net Agreed Amount.

---

[10] SIA Form of Contract, Measurement Contract, 9th Edition, cl 12(4)(e)(ii). PSSCOC Form of Contract, 2014 Edition, cl 20.1(d).

VO #	Description	Contractor's Claim ($)		QS Assessment ($)		Agreed Amount ($)		Net Agreed Amount ($)
		Addition	Omission	Addition	Omission	Addition	Omission	
1	Construct temporary carpark	58,140.00	0.00	54,600.00	0.00	54,600.00	0.00	54,600.00
2	Enlarge manholes	6,553.80	0.00	5,428.80	0.00	5,428.80	0.00	5,428.80
3	Demolish and remove existing gratings, install & modify with new grating	28,600.00	0.00	26,600.00	0.00	26,600.00	0.00	26,600.00
4	Construct access track	8,269.35	0.00	8,300.00	0.00	8,300.00	0.00	8,300.00
5	RC kerb	9,547.78	0.00	0.00	0.00	0.00	0.00	0.00
6	Construct link driveway	4,309.77	0.00	0.00	0.00	0.00	0.00	0.00
7	Reconstruct paving from 1.0m to 2.0m	14,907.66	0.00	0.00	0.00	0.00	0.00	0.00

Fig. 7.4 Summary of variation orders.

(Continued)

VO #		Contractor's Claim ($)		QS Assessment ($)		Agreed Amount ($)		Net Agreed Amount ($)
	Description	Addition	Omission	Addition	Omission	Addition	Omission	
8	Electrical and lighting Works	1,287.00	0.00	0.00	0.00	0.00	0.00	0.00
9	Additional paving	14,560.00	0.00	0.00	0.00	0.00	0.00	0.00
10	Supply only pavers	2,215.20	0.00	0.00	0.00	0.00	0.00	0.00
11	Supply only tinted pavers	1,416.80	0.00	0.00	0.00	0.00	0.00	0.00
12	Trim sand profile from the property front to the coping stone at platform	NIL	NIL					
	Total	149,807.36	0.00	94,928.80	0.00	94,928.80	0.00	94,928.80

Fig. 7.4   (Continued)

VO 1   Construction of temporary carpark

Item	Description	Contractor's Claim					QS assessment			Remarks
		Qty.	FSR FY 99/00 ref	Unit	Rate $	Amount $	Qty.	Rate $	Amount $	
	At Open Carpark Area									
1	Site clearance	8,700.00	P-2/1	m²	1.80	15,660.00	8,700.00	1.80	15,660.00	
2	Excavation, ave. depth exceed 100mm & n.e. 200mm deep	8,700.00	P-2/3	m²	0.80	6,960.00	8,700.00	0.80	6,960.00	
3	Trimming and levelling to fall	8,700.00	P-2/4	m³	0.30	2,610.00	0.00	0.45	0.00	SOR for Excavation includes trimming & levelling & therefore this item should not be paid.
4	Supply & lay 100mm thick aggregate	8,700.00	*	m²	2.50	21,750.00	8,700.00	2.50	21,750.00	

Fig. 7.5   Particulars of VO 1.

VO 1   (Continued)

Item	Description	Contractor's Claim					QS assessment			Remarks
		Qty.	FSR FY 99/00 ref	Unit	Rate $	Amount $	Qty.	Rate $	Amount $	
	At Sheltered Area									
1	Excavation, ave. depth exceed 100mm & n.e. 200mm deep	3,100.00	P-2/3	m²	0.80	2,480.00	3,100.00	0.80	2,480.00	
2	Trimming and levelling to required falls	3,100.00	P-2/4	m²	0.30	930.00	0.00	0.30	0.00	SOR for Excavation includes trimming & levelling & therefore this item should not be paid.
3	Supply & lay 100mm thick aggregate	3,100.00	*	m²	2.50	7,750.00	3,100.00	2.50	7,750.00	
	TOTAL					58,140.00			54,600.00	

Fig. 7.5   (Continued)

VO 2 Enlarging manholes including labor for hacking and reconstructing existing manholes

Item	Description	Contractor's Claim					QS assessment			Remarks
		Qty.	FSR FY 99/00 ref	Unit	Rate $	Amount $	Qty.	Rate $	Amount $	
	Existing RC slab									
1	Hacking up	19.00	P-4/7	m³	37.00	703.00	19.00	37.00	703.00	
2	Removal offsite	15.00	P-2/12	m³	14.00	210.00	15.00	14.00	210.00	
3	Backfilling & compacting	7.00	P-2/8	m³	4.00	28.00	7.00	4.00	28.00	
4	Hardcore	4.38	P-3/2	m³	28.68	125.62	4.38	28.68	125.62	
5	G20 concrete	3.50	P-1/5	m³	102.33	358.16	3.50	102.33	358.16	
6	BRC B7	43.77	P-12/3	m²	4.24	185.58	43.77	4.24	185.58	
7	Paver block	43.77	P-1A/3	m²	45.00	1,969.65	43.77	45.00	1,969.65	
8	30mm sand blinding	1.31	P-3/8	m³	27.90	36.55	1.31	27.90	36.55	

Fig. 7.6 Particulars for VO 2.

VO 2 (Continued)

Item	Description	Contractor's Claim					QS assessment			Remarks
		Qty.	FSR FY 99/00 ref	Unit	Rate $	Amount $	Qty.	Rate $	Amount $	
	Existing sluice manholes-3nos.									
1	Hacking up	4.00	P-4/7	m^3	37.00	148.00	4.00	37.00	148.00	
2	Removal offsite	3.00	P-2/12	m^3	14.00	42.00	3.00	14.00	42.00	
3	Formwork	1.92	C-11/3	m^2	20.32	39.01	1.92	20.32	39.01	
4	Extra over	1.92	C-11/4	m^2	2.70	5.18	1.92	2.70	5.18	
5	G20 concrete	2.00	P-1/5	m^3	102.33	204.66	2.00	102.33	204.66	
6	BRC B7	9.50	P-12/3	m^2	4.24	40.28	9.50	4.24	40.28	
7	Paver block	5.00	P-1A/3	m^2	45.00	225.00	5.00	0.00	0.00	Pavers are not payable as VO because already paid in contract sum.
8	30mm sand blinding	0.13	P-3/8	m^3	27.90	3.63	0.13	27.90	3.63	

	Existing drainage manholes									
1	Hacking up	4.00	P-4/7	m³	37.00	148.00	4.00	37.00	148.00	
2	Removal offsite	3.50	P-2/12	m³	14.00	49.00	3.50	14.00	49.00	
3	Formwork	11.30	C-11/3	m²	20.32	229.62	11.30	20.32	229.62	
4	Rebar	166.00	C-11/10	kg	0.75	124.50	166.00	0.75	124.50	
5	G20 concrete	3.50	P-1/5	m³	102.33	358.15	3.50	102.33	358.15	
6	Ceramic floor tile to floor	2.00	SM-15/13	m²	205.50	411.00	2.00	205.50	411.00	
7	Paver block	20.00	P-1A/3	m²	45.00	900.00	20.00	0.00	0.00	Pavers are not payable as VO because already paid in contract sum.
8	30mm sand blinding	0.33	E-3/8	m³	27.90	9.21	0.33	27.90	9.21	
	TOTAL					6,553.80			5,428.80	

Fig. 7.6 (Continued)

It would be noted in Fig. 7.4 that Contractor's claim for VO 1 to VO 12 was $149,807.36. The QS assessment was $94,928.80. Through a series of consultations and clarifications, the Contractor agreed with the QS assessment of $94,928.80 for VO 1 to VO 12.

During meetings to clarify the VO claims, it is common for the Contractor and QS to refer to the detailed quantities and rates applicable to each item so as to identify the differences. Hence, for example, the particulars of VO 1 are provided in Fig. 7.5. In the particulars for VO 1, the Contractor's measured quantities, rates and amount for each item are compared to the corresponding measured quantities, rates and amount by the QS. Where there is a difference in amount, the reason should be provided in the "Remarks" column.

In VO 1 in Fig. 7.5, the Contractor's claim was $58,140.00 and QS Assessment was $54,600.00. The QS assessed the item for 'Trimming and levelling to fall' as not payable as the rate in the Schedule of Rates (SOR) for the previous item of Excavation includes trimming and levelling to fall. This item accounts for the difference in $3,540.00 between the Contractors claim and the QS assessment. Through this exercise, all parties are able to understand the differences between the Contractor's claim and the QS assessment.

In VO 2 in Fig. 7.6, the particulars of the Contractor's claim and QS assessment are similarly compared with each other. The difference arose in a claim by the Contractor for paver blocks which was assessed by the QS to be part of the original contract works and thus not payable as a VO.

7.7 Agreed Valuation of Variations

It is advisable for parties to sign-off on the agreed valuation of variations as soon as possible. Due to lapses in memory or self-serving interest, it is not uncommon for a party to deny that there was any agreement on the valuation of variations and this may trigger a painful process for adjudication or arbitration at a later stage.

The standard form contracts do not provide that the valuation of variations should be agreed. The procedure as expressed in contracts is for the Contractor to justify and claim the amount for Variation work before the Architect or Superintending Officer certifies the payment. Prior to certification for payment, the QS will assess and recommend accordingly. If there is any dispute, it will be for the Contractor to raise the issue ultimately through adjudication or arbitration. Naturally, parties will avoid going to adjudication or arbitration, both of which will consume the time, costs and energy of the parties to varying degrees.

Thus it is sensible for the QS to clarify with the Contractor the differences in the valuation of Variations and try to bridge the gap or difference in the claims. A good example is shown in Fig. 7.4 where there was an initial difference of $54,878.56 ($149,807.36 – $ $94,928.80) from VO 1 to VO 12. The details were fleshed out and there was agreement with the Contractor in respect to the net agreed amount to claim for Variation. With agreement, it is unlikely for there to be any adjudication or arbitration in respect of the valuation of Variations.

7.8 Valuation of Variations in a Payment Claim

After carrying out the variation work, the Contractor may claim for the Variation work in a Payment Claim.[11]

An example of Variations claimed in a Payment Claim is shown in item 5 (in bold), in the next figure. For ease of reading, the Claim Amount Particulars in Fig 5.4 is reproduced below in Fig. 7.7.

The Variation Claim and Response in Fig. 7.7 and 7.9 respectively show the position prior to any agreement between the QS and the Contractor in respect of the valuation of Variations.

---

[11]SIA Measurement Contract, 9th Edition, cl 31(4)(b). PSSCOC 2014 Edition, cl 32.1(1)(a).

Item	Description	Contract Value	% Work Done	Claim	Remarks
1.0	Preliminaries	$631,000.00	67.83%	$428,020.00	See Section 1.0
2.0	Schedule of works				
2.1	Piling	$133,400.00	100%	$137,090.00	See Section 2.1
2.2	Site works	$124,600.00	100%	$124,600.00	See Section 2.2
2.3	Bungalow house	$10,070,000.00	27.71%	$2,790,000.00	See Collection
3.0	NSC for Air-conditioning and Mechanical Works	$134,000.00	48.89%	$65,500.00	See Section 3.0
4.0	NSC for Electrical Works	$78,000.00	24.36%	$19,000.00	See Section 4.0
5.0	Variations	NA		$13,225.00	See Section 5.0
	Value of work done by Main Contractor (Items 1.0, 2.1, 2.2, 2.3,5.0)			$3,492,935.00	
	Value of work done by NSC (Items 3.0, 4.0)			$84,500.00	
	Materials on site			$350,000.00	See invoice attached
	Total			$3,927,435.00	

Fig. 7.7   Variation claim in payment claim.

Less: Retention for value of work done (10% x $3,577,435.00)	($357,743.50)	
Less: Retention for materials on site (20% x $350,000.00)	($70,000.00)	($427,743.50)
Total		$3,499,691.50
Less: Previous payment certified		($3,000,000.00)
Claim Amount		**$499,691.50**

Fig. 7.7   *(Continued)*

Particulars of Variations

Item	Description	Instruction ref.	Value of variation claim	Remarks
1.	Pile cap	AI-1	$5,600.00	See
2.	Letter box	AI-2	$125.00	measure-
3.	Change to driveway	AI-3	$2,500.00	ment
4.	Change to timber flooring	AI-4	$3,500.00	sheets and rates
5.	Change to pump room	AI-5	$2,500.00	attached.
6.	Change to main entrance	AI-6	($1,000.00)	
	Total carried forward to Claim Amount Particulars		$13,225.00	

Fig. 7.8 Particulars of variation claim in the payment claim.

The detailed breakdown in the Variation claim for $13,225.00 is shown in Fig. 7.8 Particulars of Variations.

7.9 Valuation of Variations in a Payment Response

In response to the Payment Claim, the Owner would provide a Payment Response proposing to pay for the Variations in item 5 (in bold), Payment Response in Fig. 7.9.

The detailed breakdown in the response of $10,000.00 to the Variation claim is shown in Fig. 7.10 Particulars of Variation Response Amount in the Payment Response. Further, the Owner should also provide reasons for the differences and calculations in deriving the Response Amount as shown in Fig. 7.11.

7.10 Differences in Valuation of Variations in Payment Claims and Payment Responses

In Fig. 7.9, Item 5.0, the difference in the valuation of Variations between the claim of the Contractor and assessment of the QS is $3,225.00 ($13,225–$10,000.00) which may not be substantial. But, in

Item	Description	Contract Value	Claim Amount	Response Amount	Remarks
1.0	Preliminaries	$631,000.00	$428,020.00	$400,000.00	See Section 1.0
2.0	Schedule of works				
2.1	Piling	$133,400.00	$137,090.00	$137,090.00	See Section 2.1
2.2	Site works	$124,600.00	$124,600.00	$124,600.00	See Section 2.2
2.3	Bungalow house	$10,070,000.00	$2,790,000.00	$2,539,310.00	See Section 2.3
3.0	NSC for Air-conditioning and Mechanical Works	$134,000.00	$65,500.00	$55,000.00	See Section 3.0
4.0	NSC for Electrical Works	$78,000.00	$19,000.00	$15,000.00	See Section 4.0
5.0	**Variations**	NA	**$13,225.00**	**$10,000.00**	See Section 5.0
	Value of work done by Main Contractor		$3,492,935.00	$3,211,000.00	
	Value of work done by NSC		$84,500.00	$70,000.00	
	Materials on site		$350,000.00	$350,000.00	See invoice attached
	Total		$3,927,435.00	3,631,000.00	

Fig. 7.9 Response to variation claim in payment response.

Item	Description	Contract Value	Claim Amount	Response Amount	Remarks
	Less: Retention for value of work done (10% x $3,281,000)		($357,743.50)	($328,100.00)	
	Less: Retention for materials on site (20% x $350,000.00)		($70,000.00))	($70,000.00)	
	Total		$3,499,691.50	$3,232,900.00	
	Less: Previous payment certified		($3,000,000.00)	($3,000,000.00)	
	Final Amount		**$499,691.50**	**$232,900.00**	

Fig. 7.9   (*Continued*)

Particulars of Variations in Payment Response

Item	Description	Instruction ref.	Claim Amount	Response Amount	Reasons for differences
1.	Changes to pile cap	AI-1	$5,600.00	$5,600.00	
2.	Changes to letter box	AI-2	$125.00	0	Note 5/2
3.	Changes to driveway	AI-3	$2,500.00	0	Note 5/3
4.	Changes to timber flooring	AI-4	$3,500.00	$2,900.00	Note 5/4
5.	Changes to pump room	AI-5	$2,500.00	$2,500.00	
6.	Changes to main entrance	AI-6	($1,000.00)	($1,000.00)	
7.	Add shower door	AI-7	(deleted)		
	Total carried forward to Payment Response		$13,225.00	$10,000.00	

Fig. 7.10 Particulars of variation response amount in the payment response.

Note	Reasons for the differences
5/2	Changes to Letter Box not done.
5/3	Changes to driveway was defective due to failure to compact hard-core in accordance with instruction AI-3. Severe cracks appeared due to settlement. Contractor was instructed to reconstruct driveway.
5/4	Failed to polish surface of timber flooring as required in AI-4. Deducted 200m^2 × $3.00 = $600.00.

Fig. 7.11 Reasons and calculations in deriving the response amount.

other instances, it may be substantial and may be a cause for adjudication when the Claimant receives a Payment Response with a large deduction due to the lower QS assessment for Variations.

As mentioned in para 6.5 of chapter 6 in respect of Payment Claim and Payment Response, it may be advisable for the QS and Contractor to meet in order to agree on the valuation of Variations prior to submission of the Payment Claim. Substantial differences in the QS and Contractor's valuation of Variations would be avoided if the parties are able to agree on the valuation.

CHAPTER 8

Loss & Expense and Material Price Fluctuation Claim

In this chapter, we will discuss the following:

8.1 What is a Loss & Expense Claim?
8.2 How to Make a Claim for Loss & Expense in Accordance with PSSCOC Form of Contract
8.3 How to Make a Claim for Loss & Expense in Accordance with SIA Form of Contract
8.4 What is a Claim for Material Price Fluctuation?
8.5 Computation for Material Price Fluctuation

8.1 What is a Loss & Expense Claim?

The PSSCOC Form of Contract provides that the Contractor may recover Loss & Expense ("L&E") from the Owner where the ordinary progress of works has been affected by a number of stipulated events.[1] Though provided, L&E claims by Contractors are uncommon and, if claimed, are fraught with difficulties, as Employers do not take well to such claims and the claims are also hard to justify.

For example, upon commencement of works, the Superintending Officer gave an instruction to the Contractor for the construction of an additional story to a proposed 10-story building. Arising from this instruction, the Contractor was required not to proceed with works, but to wait for Consultants' new design for the substructure and superstructure works. The delay would be, say, 3 to 4 months. If the Contractor justified that the delay was due to activities on the critical path of the construction program and that the delay satisfied cl 14.2

---

[1] PSSCOC 2014 Edition, cl 22.1.

and cl 14.3, Extension of Time ("EOT") may be granted to the Contractor for:

(i) the delay in preparing the new design (say 4 months);
(ii) additional work in constructing the additional story (say 4 months).

Assuming that the EOT for (i) and (ii) is provided for 8 months, the Contractor will suffer L&E arising from the 8 months of EOT as follows:

(1) cost of management and staff;
(2) site accommodation;
(3) services and facilities, e.g. water and electricity;
(4) security and site protection;
(5) cost of unproductive labor;
(6) etc ...

The above items will not be recoverable in the Variation claim. The valuation of Variation takes into account value of additional work done by way of quantities of work and rates. Hence, the items (1) to (6) will have to be recovered by way of an L&E claim.

The SIA Measurement Contract 9th Edition does not provide for Contractor's recovery of L&E. However, it allows for adjustment to the Preliminary items.[2] In other words, where the Contractor has priced for a schedule of Preliminary items in the Contract Documents, the values for some of the Preliminary items may be adjusted in a proportionate manner in relation to the changes in quantities and time.

8.2 How to Make a Claim for Loss & Expense in Accordance with PSSCOC Form of Contract

Pursuant to cl 23.1, the Contractor shall provide a notice to claim for Loss & Expense. A simple notice is illustrated in Fig. 8.1.

By way of an example, we will compute the L&E claim arising from the SO's instruction to the Contractor for the construction of an additional story to the proposed 10-story building as mentioned in para 8.1.

---

[2]SIA Measurement Contractor, 9th Edition, cl 12(4)(a), (b).

[Contractor's Letterhead]

4 March 2013

Email & Registered Post

The Superintending Officer
[Address]
Attention: Mr. Tan Ah Kow

Dear Sir,

NOTICE FOR LOSS AND EXPENSE CLAIM FOR ENLARGEMENT OF CANAL CONSTRUCTION AND COMPLETION OF SENGKANG CANAL

We refer to the variation instruction dated 24 February 2013.

Pursuant to Clause 23.1 of the Conditions of Contract, we give notice for our claim for loss and expense in respect of the variation instruction to enlarge the canal. The consequences of the variation instruction are as follows:

  (i) additional utility charges;
 (ii) additional insurance premium to cover the extension of time;
(iii) additional security charges;
(iv) additional general labor to be employed;
 (v) abortive works including but not limited to removing of existing sheet piles;
(vi) other loss and expense items to be incurred as work proceeds.

Pursuant to cl 23.3 of the Conditions of Contract, we will substantiate the claim amount on a monthly basis as the variation work proceeds. We trust you are agreeable.

Yours faithfully,

_____

Name: _____

Designation: _____

Cc [Employer]
   [QS]

Fig. 8.1   Notice for claim for Loss & Expense.

The Contractor was required not to proceed with works, but to wait for Consultants' new design for the substructure and superstructure works. The Extension of Time ("EOT"), say, of 8 months will be granted to the Contractor for:

(i)   the 4-month extension in preparing the new design;
(ii) another 4-month extension due to the construction of the additional story.

Assuming that the PSSCOC 2014 Edition applies to the contract, the Contractor's right to recover L&E for the variation of an additional story to the proposed building project is as follows:

(iii) cl 22.1(a) — issue of an instruction for variation;

The recovery of L&E should also be in accordance with the definition of L&E:

(iv) cl1.1(q) — direct costs of labor, Plant, materials that were actually incurred
      — direct overhead cost that was actually incurred
      — plus 15% to the above costs

It would be necessary for any L&E claim to come within the scope of the above clauses.

Among other things, the Contractor will have to comply with the condition precedent cl 23.1, a condition precedent notice to be given by the Contractor to the SO within 60 days of the variation instruction.
The summary in respect of the L&E claim is in Fig. 8.2.

(A) L&E claim in respect of the 4-month extension in preparing the new design

In claiming for the various items in L&E, as shown below, it is necessary that each item comes within the definition of cl 1.1(q)(i), (ii) and (iii). An important point to note is the requirement in cl 1.1(q)(i) and (ii) for "cost of ... *actually incurred*" for items claimed. In other words, the

S/No.	Description	Amount	Remarks
(1)	Additional story		
	(i) L&E due to new design	$176,450.00	See (A)
	(ii) L&E due to additional story	$225,200.00	See (B)
	Sub-total	$401,650.00	
	Plus 15%	$60,247.50	
	Total	**$461,897.50**	

Fig. 8.2   Summary of L&E claim.

costs of such items would have been *paid out of pocket* before it may be claimed as L&E under cl 1.1(q)(i) and (ii).

It would be noted that any cost incurred by head office overheads, e.g. staff at head office advising site staff on operational and costs issues, has not been factored as an item of L&E claim below. Cl 1.1(q) (iii) has provided an additional 15% over the *cost actually incurred* in cl 1.1(q)(i) and (ii) to include head office and other administrative over-heads, financing charges etc. Thus, such head office overhead costs and other L&E items' costs would be subsumed within cl 1.1(q)(iii).

(a) Management and staff

During the 4 months of preparation for the new design, little to no work would be carried out by the Contractor. The Management and staff deployed on the site, being part of the overheads of the Contractor actually incurred on the Site (cl 1.1(q)(ii)), may be claimed as an L&E item.

As a matter of evidence, it is important to provide documentary proof of actual costs incurred in paying salaries, e.g. bank statements, notice of payment of salaries. Documentary proof of actual incurrence of cost would be applicable to all other L&E claim items.

The following is an example of the L&E claim for Management and staff due to the delay for new design:

S/No.	Staff	Time on Site	Rate ($/month)	Amount
1.	1 no. site engineer	Full-time	$2,500.00	$2,500.00
2.	1 no. quantity surveyor	Full-time	$2,500.00	$2,500.00
3.	1 no. site agent	Full-time	$2,000.00	$2,000.00
4.	1 no. site clerk	Full-time	$1,800.00	$1,800.00
5.	1 no. safety officer	Full-time	$2,500.00	$2,500.00
				$11,300.00

Hence, the L&E claim for Management and staff during the delay = 4 months × $11,300 = $45,200.00 (to collection page)

(b) Site Accommodation

The contractor would also have to pay for rental of site offices and other temporary buildings during the 4-month delay, being overheads incurred on the Site, cl 1.1(q)(ii). For example:

Description	Rate	Amount
3 no. site offices	$2,000 per month × 3 nos.	$6,000.00
3 nos. toilets	$500 per month × 3 nos.	$1,500.00
		$7,500.00

Documentary proof includes rental agreement, invoices, receipts in respect of the renting of site offices and other facilities.

Hence, the L&E claim for Site Accommodation = 4 months × $7,500.00 = $30,000.00 (to collection page).

(c) Utility services

The Contractor would also be required to pay for utility services during the 4-month period of delay, as these services are part of the overheads incurred on the Site, cl 1.1(q)(ii), as follows:

Utility expenses	Amount
SP Services bill dated 12 Jan	$1,550.00
SP Services bill dated 12 Feb	$1,800.00
SP Services bill dated 12 Mar	$1.600.00
SP Services bill dated 12 Apr	$1,400.00
SingTel bill dated 10 Jan	$500.00
SingTel bill dated 10 Feb	$550.00
SingTel bill dated 10 Mar	$520.00
SingTel bill dated 10 Apr	$600.00
Total	$8,250.00

Documentary proof includes bills and receipts.

The total claim for utility bills for the 4-month delay = $8,250.00 (to collection page).

(d) Security

Based on the contract to provide security services which are part of the overheads incurred on the Site, cl 1.1(q)(ii), the Contractor paid $5,000 per month. Hence, the L&E claim = 4 months × $5,000 per month = $20,000.00 (to collection page).

(e) Construction equipment

Many construction equipment may be hired by the Contractor for the project. Being part of the overheads incurred on the Site, cl 1,1(q)(ii), these may be claimed as follows:

Construction equipment	No.	Rate($/month)	Amount
Excavator	3	$500.00 each	$1,500.00
Dumper	5	$300.00 each	$1,500.00
Mechanical compactor	5	$200.00 each	$1,000.00
Tower crane	1	$3,000.00	$3,000.00
		Total	$7,000.00

Documentary proof includes contracts, invoices, and receipts in respect of the hire of construction equipment.

Hence, the L&E claim for Construction Equipment = 4 months × $7,000.00 = $28,000.00 (to collection page).

(f) Unproductive labor

There would also be unproductive general labor deployed on site as part of the direct costs of labor, cl 1.1(q)(i), as follows:

Ten (10) laborers × $1,000 × 4 months = $40,000.00 (to collection page).

(g) Insurance

The insurance policy for workmen injuries, public liability and damage to property may have to be extended to cover the delay period of 4 months and this is part of the overheads incurred on the Site, cl 1.1(q)(ii). Additional premium may have to be paid, say, $5,000.00 (to collection page).

Documentary proof in the extended 4-month period policy, invoice and receipts may be necessary.

(h) Collection

The summary for the L&E claim in respect of the 4-month extension in preparing the new design is as follows:

S/No.	Description	Amount
1.	Management & staff	$45,200.00
2.	Site accommodation	$30,000.00
3.	Utilities	$8,250.00
4.	Security	$20,000.00
5.	Construction equipment	$28,000.00
6.	General labor	$40,000.00
7.	Insurance	$5,000.00
	Total to Fig. 8.2	$176,450.00

Fig. 8.3 Summary of L&E items claimed for the 4-month extension in preparing new design.

(B) L&E claim in respect of the 4-month extension due to the construction of the additional story.

The items of the L&E may be similar to the 4-month extension for
preparation for new design as follows:

S/No.	Description	Amount	Remarks
1.	Management & staff	$45,200.00	No change to the value claimed earlier as we assume the same number of staff on site for the 4-month EOT.
2.	Site accommodation	$30,000.00	No change to the value claimed earlier as we assume the same site accommodation facilities.
3.	Utilities	$35,000.00	The values for utilities would be increased due to higher usage during the construction of the additional story.
4.	Security	$20,000.00	No change to the value claimed earlier as we assume the same security detail.
5.	Construction Equipment	$50,000.00	The values for Construction Equipment would be increased due to more equipment being used for the construction of the additional story.
6.	General Labor	$40,000.00	No change to the value claimed earlier as we assume the same number of general labor used during the EOT.
7.	Insurance	$5,000.00	Same as that claimed earlier.
	Total to Fig. 8.2	$225,200.00	

Fig. 8.4 Summary of L&E items claimed for the 4-month extension due to
construction of additional story.

(v) L&E Interim Account

In preparing the accounts in (i) and (ii) for submission within the 30 days of the notice of claim,[3] many numbers and items may not be available until they are actually expended progressively over the period of the 8-month Extension of Time. Nonetheless, cl 23.3 requires an L&E interim account for that particular claim. Moving forward, cl 23.3 provides, among other things, that:

(a) The Contractor shall send further interim accounts by giving the accumulated amount of the claims;
(b) The Contractor shall give any further grounds of claims;
(c) The Contractor shall send a Final Account of the claims within 30 days of the end of the particular L&E event;
(d) Actual cost incurred.

A difficult area in the computation or assessment for L&E under the PSSCOC Form of Contract is the requirement to obtain L&E costs "*actually incurred*".[4] This means costs actually paid out of pocket or sustained.

For example, in the derivation of L&E claim for utilities, one should derive the cost of utilities actually paid for the initial 4-month extension in preparing new design and the subsequent 4-month extension in the construction of the additional story. As there was no work in the initial 4-month, the utilities bills incurred in respect of this period were attributable to the instruction not to carry out work for that 4-month period. Hence, the whole of the utilities bills for the 4-month period may be claimed as actual cost incurred due to the instruction not to carry out any work.

In the derivation of L&E claim for utilities arising from the second 4-month period for the construction of the additional story, there are various challenges in deriving actual cost:

(a) The utilities' bills usually provide cost of utility for the whole of the works. Hence, part of the utility costs may be attributable to other works that are not due to the construction of the additional story. This difficulty could occur in other items of claim as well.

---

[3] PSSCOC 2014 Edition, cl 23.3.
[4] PSSCOC 2014 Edition, cl 1.1(q)(i), (ii).

(b) Lack of documentary proof in actual cost incurred.

Some of the difficulties may be overcome by proper documentation. In particular, where there is a likelihood for claim, payments must be properly recorded with details of the amount paid, time of payment, description and reasons for payment etc.

8.3 How to Make a Claim for Loss & Expense in Accordance with SIA Form of Contract

In the SIA Form of Contract, there are no provisions for the recovery of L&E. However, cl 12(4)(a) and (b) provide for payment of adjustment to preliminary items based on cl 5.

(i) L&E claim in respect of the 4-month extension in preparing the new design

Take for example a situation where items 1 to 7 below are part of the Preliminary Items in the contract documents for which the Contractor has priced and indicated "T" for adjustment to be based on time.

S/No.	Preliminary Items
1.	Management & staff
2.	Site accommodation
3.	Utilities
4.	Security
5.	Construction equipment
6.	General labor
7.	Insurance

Fig. 8.5 Summary of Preliminary Items claimed for the 4-month extension in preparing new design

The values of the above-mentioned items in the Preliminaries may be adjusted pro-rata based on the 4-month extension of time. It is to be noted as follows:

(a) Only the Preliminary items in the Contractor Documents may be adjusted for payment. Any items not in the Preliminaries may not be claimed.

(b) The amount of adjustment made to the Preliminaries is based on quantities or time, using, where reasonable, the "Make-up of Contractor's Prices under cl 5". Hence, the adjustment to Preliminaries is not based on actual cost incurred as provided expressly in the PSSCOC 2014 Edition.

8.4 What is a Claim for Material Price Fluctuation?

In some contracts, the Contractor may be able to claim for changes in prices of material, e.g. concrete and steel reinforcement. Although Material Price Fluctuation provisions are standard clauses in the SIA and PSSCOC Form of Contracts,[5] such provisions are not commonly used. For most contracts, there is no payment to Contractors for changes in material prices during the progress of the works. Hence, during tender, the Contractor may price at a certain rate for concrete and steel reinforcement and he will be paid the same amount despite a price hike in concrete and steel reinforcement midway in the construction project.

Where there is a risk of a rise in prices of basic building materials, the Contractor may have to allow for such escalation in the tender, resulting in higher tender prices. Thus, in mitigation of such a rise in tender prices, the Owner may resort to the use of the Material Price Fluctuation clause in respect of basic building materials such as concrete and steel reinforcement. By virtue of the clause, the Contractor would be able to claim for any increase in prices for concrete and steel reinforcement based on a price-linked index provided by the Building and Construction Authority.

It is expected that with the use of the clause, the Contractor may rely on this provision to claim for any increase in material prices during the contract period without increasing the tender price.

8.5 Computation for Material Price Fluctuation

The Building and Construction Authority provides price-index linked monthly indices of concrete and steel reinforcement as shown in Fig 8.6. The indices below are fictitious for purposes of illustration.

---

[5] SIA Measurement Contract, 9th Edition, cl 39. PSSCOC 2014 Edition, cl 33.

MONTHLY MATERIAL PRICE INDICES FOR IMPLEMENTATION OF
FLUCTUATION CLAUSE
[Base Period = June 2006]

Year/ Month	Concrete	Steel reinforcement
Nov 2015	130	50
Oct 2015	124	56
Sep 2015	120	60

[Indices for both concrete and steel reinforcement in June 2006 = 100]

Fig. 8.6   Monthly material price indices for purpose of Material Price Fluctuation.

As illustrated in Fig. 8.6, there is an increasing trend in concrete prices from Sep 2015 to Nov 2015, as shown by the increase in indices from 120 to 130. But there is a decreasing trend in prices for steel reinforcement for the same period with indices decreasing from 60 to 50.

In order to understand the computation for Material Price Fluctuation, let us work through an example as follows:

As shown in the invoice, the Contractor purchased concrete in Nov 2015 at, say, $90 per m³ at index 130.

In June 2006, where the index was 100,
the price of the concrete would have been = $\dfrac{90 \times 100}{130}$

$$= \$69.23 \text{ per m}^3$$

At the time of tendering in June 2014, if the index was, say, 110,
the concrete price would have been = $\dfrac{69.23 \times 110}{100}$

$$= \$76.15 \text{ per m}^3$$

Thus, the increase in concrete price from the time of tendering in June 2014 to the time of purchase in Nov 2015 would have been = $90.00 − $76.15

$$= \$13.85 \text{ per m}^3$$

The Contractor should be reimbursed for the concrete purchased in Nov 2015 at the rate of <u>$13.85 per m³</u>.

If the contract provides expressly, deduction from the payment of the Contractor due to the reduction in steel reinforcement prices may also be computed.

The adjustment for Material Price Fluctuation would be included in Payment Certificates and paid accordingly.

CHAPTER 9

Time for Completion

In this chapter, we will discuss the following:

9.1 Completion Date as stipulated in the Contract
9.2 Extension of Time
9.3 Delay
9.4 Concurrent Delay
    9.4.1 Dominant Cause
    9.4.2 Apportioned delay
    9.4.3 Malmaison Approach
    9.4.4 SIA Form of Contract on Concurrent Delay
    9.4.5 PSSCOC Form of Contract on Concurrent Delay
9.5 Issuing of Extension of Time in Practice

9.1 Completion Date as stipulated in the Contract

When a Contractor enters into a contract for the construction and completion of buildings, he is principally liable to complete the project:

(a) within the time period stipulated in the Contract;
(b) within the cost of construction tendered;
(c) in compliance with the specifications and drawings.

If the Contractor fails to complete the project by the stipulated date, the Contractor will be liable to pay Liquidated Damages to the Owner. In a more serious situation, the Contractor's failure to complete the project may also be grounds for termination of the Contractor's employment by the Owner.

Naturally, there will be events in which the Completion Date as stipulated may be extended. If that occurs, the Contractor will not be liable to pay Liquidated Damages for the period of extension.

9.2 Extension of Time

If the Contractor is unable to complete the project by the Completion Date as stipulated in the Contract, he is likely to apply for an Extension of Time. With the Extension of Time granted, the Contractor will be able to avoid paying Liquidated Damages for delay. To be eligible for Extension of Time, the Contractor must satisfy the following[1]:

(a) Any condition precedent in the Contract with respect to the notice of Extension of Time must be satisfied;
(b) The event requiring an Extension of Time must be one of the events provided in the Contract that entitles the Contractor to apply for Extension of Time;
(c) The event must be one that would delay the completion, notwith-standing reasonable steps taken to reduce or avoid the delay.

A simple time-line showing an Extension of Time is in Fig. 9.1.

A notice for Extension of Time is illustrated in Fig. 9.2.

Upon the Contractor satisfying the notice of Extension of Time, the Architect or the Superintending Officer, as the case may be, will have to consider, decide and issue the Extension of Time in accordance with the Contract.

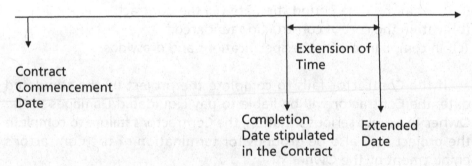

Fig. 9.1   Time-line showing extension of time.

---

[1]SIA Measurement Contract, 9th Edition, cl 23. PSSCOC 2014 Edition, cl 14.

[Contractor's Letterhead]

4 March 2013

<u>Email & Registered Post</u>

The Superintending Officer
(Address)
<u>Attention: Mr. Tan Ah Kow</u>

Dear Sir,

NOTICE FOR EXTENSION OF TIME FOR THE CONSTRUCTION AND COMPLETION OF SENGKANG CANAL

We refer to the variation instruction dated 24 February 2013.

Pursuant to Clause 14.3(1), Public Sector Standard Conditions of Contract ("PSSCOC"), we give notice that the completion date for the above-captioned project will be delayed by the Variation Instruction and we are entitled to Extension of Time under Cl 14.2(h). As required under the Cl 14.3(1) we forward as follows:

(a) Contract reference: Cl 14.2(h) – The issue of any instruction for a Variation
(b) The reasons for delay:
    (i) the additional excavation of (about 100% more earth excavated) will delay the critical path for the works;
    (ii) the existing sheet piles will be extracted and stronger sheet piles will be installed to allow deeper excavation work;
    (iii) fabrication and installation of additional reinforcement bars will be carried out;
    (iv) additional concrete casting will be carried out.
(c) The length of the delay is approx. 6 months from the contract completion date.
(d) The extension of time required is 6 months from the contract completion date.

The effect of the Variation Instruction on the program would be as shown in the revised program attached herewith showing the delay for the excavation works and subsequent activities.

We look forward to your early approval for the 6-month Extension of Time from the Contract Completion Date.

Yours faithfully

_____
[Name   ]
[Designation   ]

Cc [Employer]
   [QS]

Fig. 9.2 Notice for extension of time.

The PSSCOC Form of Contract provides that the Superintending Officer shall, in considering and deciding the Extension of Time, take into account the following[2]:

- the effect or extent of any work omitted under the Contract;
- the issue of whether the event in question is one that will delay completion of the Works;
- concurrent delays due to the default of the Contractor;
- the view that such extension is, in his opinion, fair, reasonable and necessary for the completion of the Works.

When the Superintending Officer has received sufficient information to enable him to decide on the Contractor's application and having taken into account the above, he shall, within a reasonable time, make in writing to the Contractor such Extension of Time.[3]

In the SIA Form of Contract, it is provided that when any delaying factor has ceased to operate as a delay and that it is possible for the Architect to decide on the length of Extension of Time, the Architect shall determine such period of extension at any time up to and including the issue of the Final Certificate.[4]

An example of Extension of Time issued by the Architect is in Fig. 9.3.

Whether there is a need to give a breakdown of the total Extension of Time granted depends on the Conditions of Contract. In most standard form contracts, there is no expressed requirement for a breakdown of the period of Extension of Time granted.

However, where there are numerous events necessitating an Extension of Time, it may be advisable to give a breakdown of the reasonable Extension of Time allocated for each event.

Upon completion of works, it is common for standard form contracts to provide for the issue of a Completion Certificate. An example of a Completion Certificate, mentioned in cl 17.1(1)(a), PSSCOC Form of Contract 2014, is shown in Fig. 9.4.

---

[2] PSSCOC 2014 Edition, cl 14.3(3).
[3] PSSCOC 2014 Edition, cl 14.3(3).
[4] SIA Measurement Contract 9th Edition, cl 23(3).

[Architect's letterhead]

Reference: [   ]
Date: [   ]

Best Contractor Pte. Ltd.
Blk 123, Kaki Bukit Avenue 3
Singapore 654321

Attention: Mr. Tan Ah Seng

Dear Sir,

EXTENSION OF TIME PROPOSED FOR THE CONSTRUCTION AND COMPLETION OF 4-STORY OFFICE BLOCK AT ANG MO KIO TOWN CENTRE

We refer to the above-captioned project and your request for Extension of Time dated [   ] reference [   ].

Pursuant to cl [   ] of the Conditions of Contract, we are pleased to grant an extension of time for ten (10) days from the Date of Completion as stipulated in the Contract. Consequently, the extended Contract Completion Date is [   ].

Yours faithfully

_____
Architect
cc. [Owner]
    [QS   ]

Fig. 9.3   Extension of time issued by the architect.

[Superintending Officer's letterhead]

Ref:[   ]
Date: [   ]   <u>Email & Registered Post</u>

Best Contractor Pte. Ltd.
Blk 123, Kaki Bukit Avenue 3
Singapore 654321

<u>Attention: Mr. Tan Ah Seng</u>

Dear Sir,

CERTIFICATE OF SUBSTANTIAL COMPLETION FOR THE PROPOSED CONSTRUCTION AND COMPLETION OF 4-STORY OFFICE BLOCK AT ANG MO KIO TOWN CENTRE

We refer to the above-captioned project.

In accordance with cl 17.1(1)(a) of the Public Sector Standard Conditions of Contract for Construction Works 2014 ("PSSCOC"), I hereby issue the Certificate of Substantial Completion in respect of the Proposed Construction and Completion of a 4-story Office Block at Ang Mo Kio Town Centre.

In my opinion, the works were substantially completed in accordance with the Contract on 1 March 2016. The Defects Liability Period shall commence on 1 March 2016 and expire on 28 February 2017.

In accordance with cl 18.1, PSSCOC, please complete the outstanding and defective works as listed in attachment by [   ].

Yours faithfully

_____
[Name   ]
Superintending Officer

Cc [Employer]
   [QS   ]

Fig. 9.4   Certificate of substantial completion.

Naturally, the Contractor would try his best to complete the whole of the works and obtain the Certificate of Substantial Completion (in the case of PSSCOC Form of Contract) or Completion Certificate (in the case of SIA Form of Contract) before the extended Contract Completion Date.

If the completion date as certified falls after the extended Contract Completion Date, Liquidated Damages may be imposed against the Contractor as provided in the Contract.

9.3 Delay

The Contractor is said to be in Delay if he fails to complete the whole of the works by the extended Contract Completion Date or Completion Date as stipulated in the Contract where no Extension of Time is granted. In such an instance, the documentary evidence would show that the date of certified completion would be a date after the extended Contract Completion Date. Where no Extension of Time is granted, the date of certified completion would be a date after the Completion Date, as stipulated in the Contract.

In the SIA Form of Contract,[5] a Delay Certificate must be issued when the Contractor is in Delay. If not, the Employer may not be able to recover the Liquidated Damages from the Contractor.[6] Examples of Delay Certificates are shown in Fig. 9.5 and Fig. 9.6.

Among other things, cl 24(2) SIA Form of Contract provides that the Employer may, but shall not be bound to, deduct Liquidated Damages upon receipt of the Delay Certificate. Consequently, no deduction is made in QS' interim valuation for payment in respect of Liquidated Damages.[7] The Employer has the discretion whether to deduct such Liquidated Damages from his payment to the Contractor at any time. If deducted by the Employer, the amount so deducted will be recorded in subsequent certificates of the Architect.

---

[5] SIA Measurement Contract, 9th Edition, cl 24(1), (2).
[6] Lian Soon Construction Pte Ltd v Guan Qian Realty Pte Ltd [1990] 3SLR(R); [1999] SGHC 25.
[7] SIA Measurement Contract, 9th Edition, cl 31(4)(h).

Delay Certificate

Date of issue: [   ]

To: Alpha Realty Pte. Ltd.
       [Address]

       Attention: [Mr. _____]

Title of Project: [   ]
Contract No: [   ]
Contractor: [   ]

Pursuant to clause 24(1) of the Conditions of Contract, I/we hereby certify that the Project has been delayed and that the Contractor was in default in not completing the Works by the Extended Contract Completion Date on [date].

As required under clause 24(1), I/we provide the following information:
(a) Contract Completion Date: [   ]
(b) Total period of Extension of Time: [   ]
(c) Extended Contract Completion Date: [   ]
(d) Date of Completion as certified: [   ]

Pursuant to clause 24(2), the Employer is entitled to recover Liquidated Damages from the Contractor at the rate of [$] per day from the Extended Contract Completion Date of   [   ] to the Date of Completion on [   ] as certified in the Completion Certificate. The total number of days in delay was [   ] days. Such Liquidated Damages may also be deducted from any payment of the Employer to the Contractor under the Contract.

_____
[Name   ]
Architect

Cc: Contractor
     QS

Fig. 9.5   Delay certificate 1.

Delay Certificate (Clause 24(1))

To the Employer: [   ]
Address: [   ]
Attention: [Mr. _____]

Project: [   ]

In accordance with clause 24(1) of the Conditions of Contract, the latest Date for Completion of the Works pursuant to Clause 22.1 of the Conditions has passed and there were no other matters entitling the Contractor to an extension of time. The Works have nevertheless remained incomplete at the latest Date for Completion. I hereby issue this Delay Certificate.

The Contract Completion Date was [   ]. In accordance with clause 23 of the Conditions, the total period of extension of time granted to the Contractor was [   ] days. Consequentially, the extended Contract Completion Date was [   ].

I certify that the Contractor is and has been in default in not having completed the Works by the extended Contract Completion Date.

I attach the Completion Certificate dated [   ] certifying that the Works have been completed on [   ] and appeared to comply with the Contract in all respects except the minor outstanding works listed in the Appendices to the Completion Certificate. Accordingly, there was a delay of [   ] days by the Contractor from the extended Contract Completion Date to the Completion Date as certified.

In accordance with clause 24(2) of the Conditions, the Employer shall be entitled to recover from the Contractor Liquidated Damages (including deduction from any payment to the Contractor) calculated at the rate of [$] per day for [   ] days delay.

Certified on: [   ]

Certified by: _____
                  [Name   ]
                  Architect

Copy to: [Contractor   ]
              [QS   ]

Fig. 9.6   Delay certificate 2.

In the PSSCOC Form of Contract, there is no requirement for a Delay Certificate. When the Works have not been completed within the Time for Completion or any extended time, the Contractor shall pay the Employer Liquidated Damages.[8]

9.4 Concurrent Delay

A Concurrent Delay refers to different causes of Delay overlapping over a period of time. Where there were a few overlapping causes of Delay, we have to determine the cause of each Delay, whether it was a Contractor-caused Delay or a non-Contractor-caused Delay.

A non-Contractor-caused Delay may be due to the Owner. For example, the Owner wishes to have additional facilities to the construction and therefore the Architect issues an instruction for the construction of the additional facilities. This may result in Extension of Time to the Contractor. The non-Contractor-caused Delay may also be due to other factors not caused by the Owner, e.g. adverse weather.

Thus non-Contractor-caused Delay may be Owner-caused Delay or other Delay not due to the Contractor. Events of non-Contractor-caused Delay are usually stipulated in the Contract as an event that may entitle the Contractor to an Extension of Time.[9] In determining whether an event is a Contractor-caused Delay or non-Contractor-caused Delay may help us in determining Extension of Time and consequently Delay attracting payment of Liquidated Damages.

Fig. 9.7 shows an example of Concurrent Delay. Construction activities A, B, C and D are all construction activities along a time-line. However, Construction activities B and D must be completed before Construction activity C begins. However, Construction activity B was delayed by the Owner and Construction activity D was delayed by the Contractor. The delays caused Construction activity C to be delayed by 10 Workdays. Ultimately, the delay had a knock-on effect of delaying the Completion Date by the same 10 Workdays.

---

[8] PSSCOC 2014 Edition, cl 16.1(1).
[9] SIA Measurement Contract, 9th Edition, cl 23(1)(a) – (q). PSSCOC 2014 Edition, cl 14(2)(a) – (q).

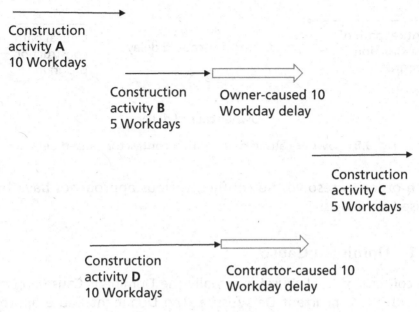

Fig. 9.7   Concurrent delay.

The question to be answered is who is responsible for the Delay? The answer to this question may also help us determine whether Extension of Time should be given to the Contractor. If not, the Contractor would be in Delay and Liquidated Damages may be payable by the Contractor.

Concurrent Delays may occur in various manners. Consider the following example:

In Fig. 9.8, there was a Contractor-caused Delay, say, severe shortage of labor on site, along the critical path program that resulted in delaying the project. During the severe shortage of labor, there was an Owner-caused Delay, say, a 7-day stop work order requested by the Owner for carrying out a marketing launch for the development. Hence, there is a Concurrent delay: one caused by the Contractor and the other by the Owner. But the Owner will convincingly argue that no Extension of Time ought to be granted to the Contractor as the Contractor-caused delay will subsist even without the marketing launch. Hence, according to the Owner, the real cause of the delay was due to Contractor's severe shortage of labor, not the Owner's marketing launch.

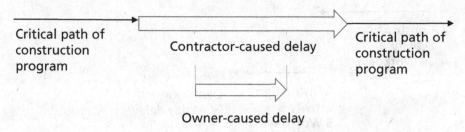

Fig. 9.8   Owner-caused delay within contractor-caused delay.

In order to resolve the conflict, various approaches have been devised as follows:

9.4.1   Dominant Cause

The concept of 'real cause' is actually the Dominant Cause approach in resolving Concurrent Delay issue. The Dominant Cause approach suggests that when a Concurrent Delay occurs, i.e., Owner-risk Delay and Contractor-risk Delay, one must decide which is the Dominant Cause that prevails over the other cause.[10] If in Fig. 9.8, the Dominant Cause for the Delay was the shortage of labor, then, notwithstanding the Owner's instruction to stop work for 7 days, there would be no Extension of Time to the Contractor.

Oftentimes, it is difficult to decide which cause is the Dominant Cause of Delay and therefore there are challenges in applying the Dominant Cause approach.

9.4.2 Apportioned Delay

Another approach is to apportion delay between the Owner and Contractor, when there was Concurrent Delay caused by both of them. This approach considers the relative importance and degree of responsibility for each delay.

In the City Inn[11] case, the Court, by apportioning delay due to the Owner and Contractor, allowed a 9-week Extension of Time out of a claim of 11 weeks Delay.

---

[10] H. Fairweather & Co Ltd v London Borough of Wandsworth (1987) 39 BLR 106 (OR).
[11] *City Inn Ltd v Shepherd Construction Ltd* [2007] CSOH 190, [2008] BLR 269, (2008)

9.4.3   Malmaison Approach

In the Malmaison Approach,[12] the Court decides on two Concurrent causes of Delay: a Relevant Event (an event eligible for Extension of Time under the Contract) and a non-Relevant Event. The Relevant Event which has caused the Delay is taken into account for Extension of Time, and the Concurrent non-Relevant Event is not taken into account at all.

The Malmaison Approach is harsh to Owners as a delay caused by a concurrent non-Relevant Event due to the Contractor is ignored in the granting of Extension of Time. This appears to be the accepted approach in UK.

9.4.4 SIA Form of Contract on Concurrent Delay

In light of the different approaches in resolving Concurrent Delay, we will examine the clauses in the SIA Form of Contract[13] to seek guidance on the issue.

> *Cl 24(3)(a) If while the Contractor is continuing work subsequent to the issue of a Delay Certificate, the Architect gives instructions or matters occur which would entitle the Contractor to an extension of time under Clauses 23(1)(f), 23(1)(g), 23(1)(h), 23(1)(i), 23(1)(j), 23(1)(k), 23(1)(n), 23(1)(o) and 23(1)(p) of these Conditions, and if such matters would have entitled the Contractor to an extension of time <u>regardless of the Contractor's own delay</u> and were not caused by any breach of contract by the Contractor, the Architect shall as soon as possible grant to the Contractor the appropriate further extension of time in a certificate known as a "Termination of Delay Certificate".*

> *Cl 24(3)(b)   Such further extension of time granted shall have no immediate effect nor shall it prevent the deduction or recovery of liquidated damages by the Employer until the issuance of the Termination of Delay Certificate. The Termination of Delay Certificate ... while not preventing the deduction or recovery of liquidated damages accrued up to its issuance, shall prevent the accumulation of liquidated damages during the period of the further extension of time granted.*

---

24 Constr LR 590.

[12] *Henry Boot Construction (UK) Ltd v Malmaison Hotel (Manchester) Ltd,* (1999) 70 Con LR 32 (TCC).

[13] SIA Measurement Contract, 9th Edition.

Cl 24(3)(c)   Thereafter, if the Contractor fails to proceed to complete the Works with due diligence within the period of the further extension of time granted under the Termination of Delay Certificate, the Architect shall issue a further delay certificate certifying that the Works have remained incomplete and that he is again in default in not so completing. Such certificate shall be known as a "Further Delay Certificate"... . Liquidated Damages shall re-commence accruing in favour of and be recoverable or deductible by the Employer from the issuance of the Further Delay Certificate.

The above clauses may be illustrated on a time-line in Fig. 9.9.

Fig. 9.9 illustrates cl 24(3) on Further Extension of Time on a Time-line. During the period of Delay in Fig. 9.9, the Architect gives instruction which would entitle the Contractor to EOT.

"...and if such matters would have entitled the Contractor to an extension of time <u>regardless of the Contractor's own delay</u> and were not caused by any breach of contract by the Contractor, the <u>Architect shall as soon as possible grant to the Contractor the appropriate further Extension of Time</u> ...".

The expressed words of cl 24(3)(a) above provide that the appropriate further Extension of Time should be granted notwithstanding Contractor's own delay.

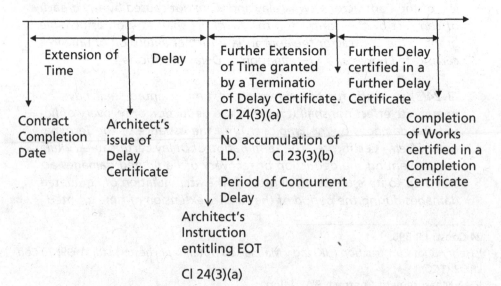

Fig. 9.9   Time-line for further extension of time.

In cl 24(3)(c), the clause provides:

"... if the Contractor <u>fails to proceed to complete the Works with due</u> <u>diligence within the period of the further Extension of Time</u>, granted under the Termination of Delay Certificate, the Architect shall issue a Further Delay Certificate ..."

The above provides for the situation where the Contractor fails to proceed with due diligence within the period of the Further Extension of Time. In such a situation, the Architect shall issue a "Further Delay Certificate" for the Further Delay as shown in Fig. 9.9.

By cl 24(3)(a), (b) and (c), the provisions require the issue of Further Extension of Time so long as it satisfies the Relevant Events,[14] notwithstanding that there were Concurrent Delays caused by the Contractor. The Conditions are silent on how the Architect is to determine the length of the Extension of Time if there was concurrency of Delay. In other words, the Conditions are silent on whether the Extension of Time is to be apportioned based on Contractor-caused or Owner-caused delay as mentioned in para 9.4.2, or the Malmaison Approach as mentioned in para 9.4.3. The Author is not aware of any decided case in point, and consequently, the matter remains, in the Author's view, within the discretion of the Architect.

9.4.5 PSSCOC Form of Contract on Concurrent Delay

Clause 14.3(3) provides:

"When the Superintending Officer has received sufficient information to enable him to decide the Contractor's application, he shall within a reasonable time, make in writing to the Contractor such Extension of Time...<u>The Superintending Officer shall also take into account any</u> <u>delays which may operate concurrently with the delay due to the</u> <u>event or events in question and which are due to acts or default on</u> <u>the part of the Contractor.</u>"

The above clause refers to the situation where there was an application for Extension of Time from the Contractor, following from cl

---

[14]SIA Measurement Contract 9th Edition, Cl 23(1)(f), 23(1)(g), 23(1)(h), 23(1)(i), 23(1)(j), 23(1)(k), 23(1)(n), 23(1)(o), 23(1)(p).

14.3(1). In making the Extension of Time to the Contractor, the Superintending Officer must also take into account the Delays caused by the Contractor that were concurrent with the Delays due to the Relevant Event. It *suggests* that where there are causes due to acts of default of the Contractor, the Extension of Time to be granted may be reduced in time so as to ascribe part of the Delay due to the Contractor's fault.

Clause 16.4 provides:

"...if the Contractor shall fail to complete the Works or any phase or part of the Works by the Time for Completion and the execution of the Works is thereafter delayed by any of the events set out in Clause 14(2)(g) to (q) inclusive, the Employer's right to Liquidated Damages shall not be affected but...the Superintending Officer shall grant an Extension of Time pursuant to Clause 14. Such extension of time shall be added to the Time for Completion of the Works..."

The above clause refers to a situation where in the midst of a Contractor-caused Delay, the Works were concurrently delayed by any cause set out in the Relevant Events[15] (see Fig. 9.10). In such a situation, the Superintending Officer shall grant an Extension of Time pursuant to Clause 14, as shown in the time-line in Fig. 9.10.

As mentioned above, Clause 14.3(3) requires taking into account any delays due to the acts or default of the Contractor. As to how and in what manner the Superintending Officer may take into account such delays is far from clear in the provision. In such a situation, the Superintending Officer in the PSSCOC Form of Contract, in the Author's view, may exercise his discretion in the matter.

9.5 Issuing of Extension of Time in Practice

There is usually no contention by the Contractor for Extension of Time if Liquidated Damages are not imposed. An example is where the Contractor is able to complete the Works before the Contract Completion Date stipulated in the Contract. Even if no Extension of Time is granted for adverse weather or additional works carried out by

---

[15] PSSCOC 2014 Edition, cl 14.2(g) to (q).

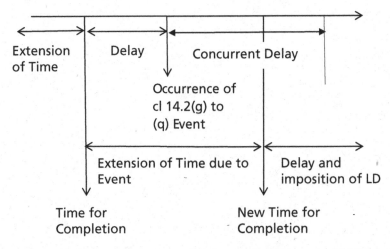

Fig. 9.10   Time-line for extension of time during delay in PSSCOC form of contract.

the Contractor, there is no relevance in any Extension of Time as there are no Liquidated Damages to be imposed.

The Extension of Time becomes relevant to the Contractor if the period of Extension of Time is insufficient. Consequently, the Contractor is in Delay and Liquidated Damages may be imposed. In such a circumstance, the Architect would usually meet with the Contractor to explain his decision, though such a meeting is not required in the Conditions of Contract. The objective for such an explanation is to show reasonableness on the part of the Architect in the assessment for Extension of Time and to show that the Delay was due to the Contractor's fault for which they would be liable to Liquidated Damages. If the objective is not met, there may be a dispute leading to arbitration or litigation, a course which may not be in the interest of either party and for which both parties would like to avoid.

CHAPTER 10

Defects in Construction Works

In this chapter, we will discuss the following:

10.1 What are Defects?
10.2 SIA Form of Contract in Resolving Defects
10.3 SIA Form of Contract in Resolving Defects During and Post Maintenance Period
10.4 PSSCOC Form of Contract in Resolving Defects
10.5 PSSCOC Form of Contract in Resolving Defects During Defects Liability Period
10.6 Powers in Resolving Defects During Progress of Works and Defects Liability Period in PSSCOC Form of Contract

10.1 What are Defects?

In the SIA Form of Contract, there is no definition for Defects. However, cl 11(3) provides as follows:

> "...the Architect may give directions for the removal or demolition of any work, goods or materials, whether fixed or unfixed, which are not in accordance with the Contract, and for their reconstruction or replacement in exact accordance with the Contract. Provided that the Architect may, but shall not be bound to, accept any work containing defects unremedied and without removal or replacement..."

Based on the above clause, Defects would refer to work, goods or materials which are not in accordance with the Contract.

In the PSSCOC 2014 Edition, cl 1.1(j) defines Defect as follows:

> ""Defect" means any part of the Works not executed, provided or completed in accordance with the Contract. For the avoidance of doubt and without limiting the generality of the expression, the term

shall be taken to include any item of plant, material, goods or work incorporated or used in the Works which does not or may not conform to the relevant quality standards or pass the test prescribed in or to be inferred from the Contract."

Based on the above explanation or definition, it is important to specify in the Contract the types of material, workmanship, standard in quality and performance of the Works. If the material or Works do not conform to the requirement specified in the Contract, it is a Defect.

Examples of specifications for materials and workmanship are as follows:

Plaster/rendering Material

(A) Cement
  Portland cement: White cement to SS 26.
(B) Sand
  To conform to BS EN 44146.
[Etc...]

Workmanship for plastering work

(C) Plastering work shall be complete to full vertical height of wall from floor to underside of concrete slab, beams or otherwise shown in drawing.
(D) All finished plaster surfaces shall be plumb, level, true, finished smooth to profiles shown on drawing, without trowel marks, cracks or other defects.
(E) All angles, intersections and corners shall be clean and accurately formed.
(F) All flat surfaces shall be level and true. Curved and other shapes and profiles shall be true to profile as shown in drawings.
[Etc...]

Tile and cement mortar material

(G) The ceramic tile shall conform to SS 675.
(H) The stone tiles shall conform to SS789.
 (I) Sand shall be clean and comply with BS EN 44146.
 (J) Cement shall comply with SS 26.
[Etc...]

Workmanship for tiling work

(K) Joints shall be true, continuous or in accordance with pattern as shown in drawings.

(L) Primer shall be applied in accordance with manufacturer's instructions.

(M) Where required, tiles shall be neatly and accurately cut.

[Etc...]

If the Contractor's materials or Works do not conform to the specifications, there is a Defect in the materials or Works for which the Contractor is liable to make good.

10.2 SIA Form of Contract[1] in Rectifying or Resolving Defects

The time-line for rectifying and resolving Defects is as follows:

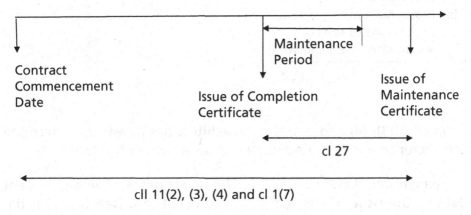

Fig. 10.1 Time-line for rectifying and resolving defects under the SIA form of contract.

During the period from the Contract Commencement Date to the issue of the Maintenance Certificate (see cl 27(3)), the Architect's power for rectifying and resolving Defects is provided in cll 11(2), (3), (4) and cl 1(7). See Fig. 10.1.

---

[1]SIA Measurement Contract, 9th Edition.

In addition, during the Maintenance Period and up to the issue of the Maintenance Certificate, the Architect has additional power under cl 27 for rectifying and resolving Defects. See Fig. 10.1.

In cl 11(2), the Architect has power to require the Contractor to *investigate* for Defects as follows:

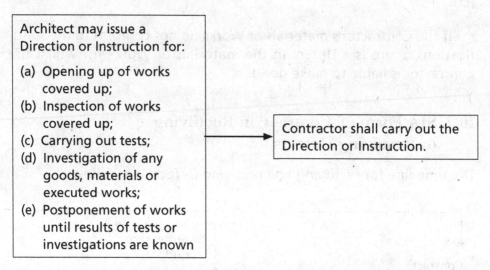

Fig. 10.2   Investigation of defects.

In cll 11(3), (4) and cl 1(7), the Architect has power to require the Contractor to rectify or resolve Defects as shown in Fig. 10.3.

As mentioned, during the period from the Contract Commencement Date to the issue of the Maintenance Certificate (see cl 27(3)), the Architect's power for rectifying and resolving Defects is provided in cll 11(2), (3), (4) and cl 1(7) as summarized in Fig. 10.2 and Fig. 10.3.

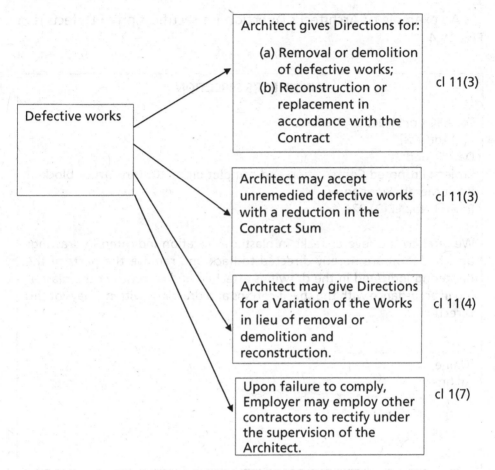

Fig. 10.3 Rectifying and resolving defects.

An example of Architect's Direction for rectification of Defects is in Fig. 10.4.

ARCHITECT'S DIRECTION

To: ABC Construction Pte. Ltd.
   [Address]
Date Issued: [   ]
Project: Proposed Construction and completion of 10-story Office block at Bukit Merah, Singapore.
Project No.: LE(D) 924

We refer to the severe cracks in plaster at location indicated in drawings attached. You are hereby directed to hack and remove the parts of the plastering indicated in the drawings attached and reconstruct the plastering in accordance with the specifications and drawings within 7 days of this Direction.

_____
[Name   ]
Architect

Cc [Employer]
   [QS]

I/We acknowledge receipt of the Architect's Direction as stipulated above and return a duplicate copy.

_____
[Name   ]
For and on behalf of the Contractor

Date:

Fig. 10.4   Architect's direction for rectification of defects.

10.3 SIA Form of Contract in Resolving Defects During and Post Maintenance Period

Cl 27 provides power to the Architect to require the Contractor to rectify and resolve Defects during and post Maintenance Period. The provisions in cl 27 do not apply to rectification of Defects during the Contract Period. Powers to rectify and resolve Defects during the Contract Period are provided in cl 11 and cl 1(7).

However, the powers to rectify and resolve Defects as provided in cll 11 (2), (3), (4) and cl 1(7) are not only applicable during the Contract Period but also during the Maintenance Period and anytime until the issue of the Maintenance Certificate. See cl 27(3).

In cl 27, the time-line for the resolution of Defects is as follows:

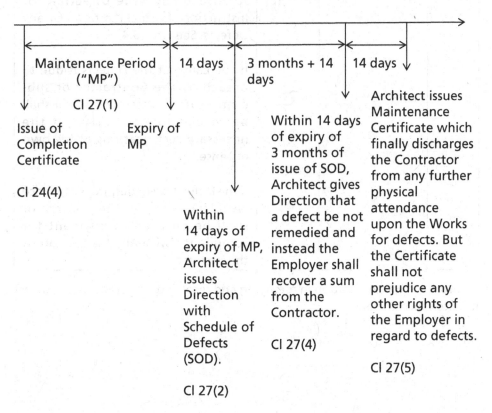

Fig. 10.5 Time-line for resolution of Defects during and post maintenance period.

Brief explanation of Fig. 10.5 is provided in Fig. 10.6.

S/No.	Event	Clause	Brief explanation
1.	Issue of Completion Certificate	24(4), 24(5)	Architect issues the Completion Certificate to the Contractor. See Fig. 10.7 and 10.8. Contractor may be required to provide an Undertaking in writing to complete Outstanding Works. See Fig. 10.9.
2.	Maintenance Period	27(1)(a)	Contractor shall complete Outstanding Works in accordance with the terms of the Completion Certificate.
		27(1)(b)	Architect may issue Directions or Instructions for making good of any Defects. See Fig. 10.4. If the cause of the Defects is due to breach by the Contractor or sub-contractors, then the Contractor shall be responsible to carry out the necessary rectifications at his own expense. If the Defects occur despite compliance with the Contract, the Contractor shall be entitled to payment for compliance with any Instruction of the Architect.

Fig. 10.6   Brief explanation in rectifying and resolving defects during and post maintenance period.

S/No.	Event	Clause	Brief explanation
3.	Expiry of Maintenance Period	27(2)	Within 14 days of expiry of the Maintenance Period, the Architect issues a Direction with Schedule of Defects specifying remaining defects. Upon receipt of the Direction, the Contractor shall repair and make good on the same terms as cl 27(1)(b). See Fig. 10.10 and 10.11.
4.	Expiry of 3 months of issue of the Schedule of Defects	27(4)	Within 14 days of the expiry of 3 months of the issue of Schedule of Defects, the Architect gives a Direction that the Defects be not remedied and instead the Employer may deduct from any monies due to the Contractor or recover from the Contractor the cost to engage other contractors to rectify the defects. See Fig. 10.12.
5.	Issue of Maintenance Certificate	27(5)	After the Defects have been resolved as above, the Architect issues a Maintenance Certificate to discharge the Contractor from any further physical attendance upon the Works. See Fig. 10.13.

Fig. 10.6   (*Continued*)

COMPLETION CERTIFICATE

Ref:[   ]
Date: 1 March 2016

Best Contractor Pte. Ltd.
Blk 123, Kaki Bukit Avenue 3
Singapore 654321

Attention: Mr. Tan Ah Seng

PROPOSED CONSTRUCTION AND COMPLETION OF 4-STORY OFFICE BLOCK AT ANG MO KIO TOWN CENTRE

In accordance with cl 24(4) of the Conditions of Contract, I hereby issue the Completion Certificate as the Works appear to be complete and to comply with the Contract in all respects on **1 March 2016** (except for the outstanding works listed in the Schedule to this Certificate.).

The Maintenance Period of the Works shall commence on 1 March 2016 and will expire on 28 February 2017.

Certified by:

_____
[Name   ]
Architect

Cc [Employer]
       [QS]
       [CE]
       [ME]
       [EE]

Fig. 10.7   Completion certificate.

Schedule to the Completion Certificate

PROPOSED CONSTRUCTION AND COMPLETION OF 4-STORY OFFICE BLOCK AT ANG MO KIO TOWN CENTRE

<u>Outstanding Works:</u>

(1) Plot 1: timber trellis at car park has not been completed.
(2) Gate control switch at main gate has not been completed.
(3) House no. has not been completed and displayed.
(4) Defects issued in Direction dated [   ] have not been rectified.
(5) ME operations manual has not been submitted.
(6) As-built drawing has not been submitted.
(7) Landscaping Works have not been completed.
(8) Painting of internal doors has not been completed

As provided in an Undertaking in writing dated 25 February 2016 by the Contractor, the Outstanding Works shall be completed by <u>1 April 2016.</u>

Fig. 10.8   Schedule of outstanding works to the completion certificate.

[Contractor's Letterhead]

Date: 25 February 2016

[Name of Employer]
[Address]

<u>Attention: Mr. Tan Ah Huat</u>

UNDERTAKING TO COMPLETE OUTSTANDING WORKS
PROPOSED CONSTRUCTION AND COMPLETION OF 4-STORY OFFICE BLOCK
AT ANG MO KIO TOWN CENTRE

Pursuant to cl 24(5) of the Conditions of Contract, I/We hereby undertake to complete all Outstanding Works by 1 April 2016. The Outstanding Works are as follows:

(1) Plot 1: timber trellis at car park has not been completed.
(2) Gate control switch at Main Gate has not been completed.
(3) House no. has not been completed and displayed.
(4) Defects issued in Direction dated [   ] have not been rectified.
(5) ME operations manual has not been submitted.
(6) As-built drawing has not been submitted.
(7) Landscaping works have not been completed.
(8) Painting of internal doors has not been completed.

_____
[Name of Authorized Person]
[Designation]

Fig. 10.9 Contractor's undertaking in writing to complete outstanding works.

ARCHITECT'S DIRECTION

To: ABC Construction Pte. Ltd.
     [Address]
Date Issued: [   ]

Schedule of Defects to Proposed Construction and Completion of 10-story Office Block at Bukit Merah, Singapore.
Project No.: LE(D) 924

The Maintenance Period has expired on [   ], but there are still defects not rectified. In accordance with cl 27(2) of the Conditions of Contract, I forward the Schedule of Defects attached herein for your rectification and making good.

If you fail to rectify the Defects within 3 months of this Direction, we reserve our rights in the Contract and other actions against you.

_____
[Name   ]
Architect

Cc   [Employer]
      [QS]

_____

I/We acknowledge receipt of the Architect's Direction as stipulated above and return a duplicate copy.

_____
[Name   ]
For and on behalf of the Contractor

Date:

Fig. 10.10   Direction to rectify accompanied with schedule of defects.

Schedule of Defects

S/No.	Description	Location
1.	Damaged toilet doors	Toilets 1-1, 2-1, 3-1
2.	Defective door closers	Rooms 1-4, 1-5, 3-5
3.	Cracks to plaster walls	Corridors of story 1, meeting room 2-5
4.	Leaks in water pipes	Toilets 1-1, 2-1, 3-1
5.	Missing floor trap cover	Toilet 1-1
6.	Missing lighting fitting	Room 1-4

Fig. 10.11 Schedule of defects.

ARCHITECT'S DIRECTION

To: ABC Construction Pte. Ltd.
      [Address]
Date Issued: [   ]

Recovery of Sum due to Defects in Proposed Construction and completion of 10-story Office Block at Bukit Merah, Singapore.
Project No.: LE(D) 924

To date, the Defects listed in attachment are still not rectified by you. The Defects were part of the Schedule of Defects forwarded to you on [   ] for making good more than 3 months ago.

Pursuant to cl 27(4) of the Conditions of Contract, I hereby direct that the Defects in attachment be not remedied and instead the Employer may deduct from any monies otherwise due to you or recover such sums from you. The Quantity Surveyor will assess the estimated cost which the Employer would incur in having to employ other contractors to make good and all other related costs and the Employer may deduct or recover such sums from you.

_____
[Name   ]
Architect

Cc   [Employer]
        [QS]

I/We acknowledge receipt of the Architect's Direction as stipulated above and return a duplicate copy.

_____
[Name   ]
For and on behalf of the Contractor

Date:

Fig. 10.12 Direction for recovery of sum instead of rectifying of defects.

Maintenance Certificate

To: ABC Construction Pte. Ltd.
     [Address]
     Date Issued: [   ]

Project: Proposed Construction and Completion of 10-story Office Block at Bukit Merah, Singapore
Project No.: LE(D) 924

Pursuant to cl 27(5) of the Conditions of Contract and as all defects notified by the Architect to the Contractor have been made good or dealt with, I hereby issue the Maintenance Certificate.

Certified by:

[Name   ]
Architect

cc. [Employer]
     [QS]
     [CE]
     [ME]
     [EE]

Fig. 10.13   Maintenance certificate.

10.4 PSSCOC Form of Contract in Resolving Defects

The time-line for the rectifying and resolving of Defects is as follows:

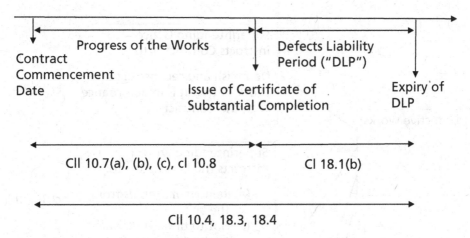

Fig. 10.14   Time-line for rectifying and resolving defect under the PSSCOC form of contract.

During the progress of the Works, cll 10.7(a), (b), (c) and cl 10.8 provide for Superintending Officer's powers to rectify and resolve Defects. See Fig. 10.14.

During DLP, the Superintending Officer's power to rectify and resolve Defects is provided in cl 18.1(b). See Fig. 10.14.

During the progress of the Works and DLP, the Superintending Officer has power under cll 10.4, 18.3 and 18.4 in respect of rectifying and resolving Defects. See Fig. 10.14.

In cll 10.7 and 10.8, the Superintending Officer has power, *during the progress of the Works*, to require the Contractor to rectify or resolve Defects as follows:

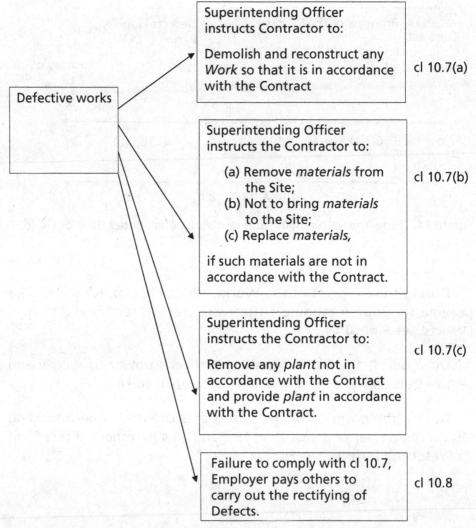

Fig. 10.15　Rectifying and resolving defects during the progress of works under the PSSCOC form.

It is noted that the power to rectify Defects as provided in cll 10.7 and 10.8 is to be exercised during the *progress of the Works* and not during the Defects Liability Period.

The Superintending Officer's power to require rectification during the Defects Liability Period is in cl 18.

It is further noted that the powers in cl 10.4, cl 18.3 and cl 18.4 would apply to Defects during the progress of the Works and Defects Liability Period.

10.5 PSSCOC Form of Contract in Resolving Defects During Defects Liability Period

The time-line for the rectification and resolution of Defects in the PSSCOC Form during Defects Liability Period is in Fig. 10.16.

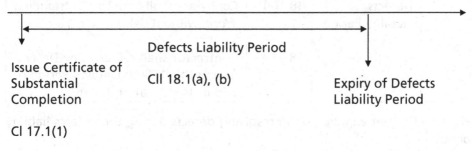

Fig. 10.16   Time-line for resolution of defects during the defects liability period.

A brief explanation of Fig. 10.16 is provided in Fig. 10.17.

S/No.	Event	Clause	Brief Explanation
1.	Issue of Certificate of Substantial Completion	17.1(1)	Superintending Officer issues the Certificate of Substantial Completion to the Contractor. See Fig. 10.18 and Fig. 10.19. Contractor may give notice of completion accompanied with an Undertaking in writing to complete Outstanding Works during the Defects Liability Period. See Fig. 10.9 with necessary modifications for application to PSSCOC Form of Contract.
2.	Defects Liability Period	18.1(a)	Contractor shall complete Outstanding Works without delay.
		18.1(b)	Contractor shall remedy any Defects during the Defects Liability Period or within 14 days after its expiry.

Fig. 10.17   Brief explanation in resolving defects during the defects liability period.

CERTIFICATE OF SUBSTANTIAL COMPLETION

Ref: [   ]

Date: 1 March 2016

Best Contractor Pte. Ltd.
Blk 123, Kaki Bukit Avenue 3
Singapore 654321

Attention: Mr. Tan Ah Seng

PROPOSED CONSTRUCTION AND COMPLETION OF 4-STORY OFFICE BLOCK AT ANG MO KIO TOWN CENTRE

In accordance with cl 17.1(1)(a) of the Conditions of Contract, I hereby issue the Certificate of Substantial Completion for the whole of the Works on **1 March 2016** (except for the outstanding works listed in the Schedule to this Certificate.). In my opinion, the Works were substantially completed in accordance with the Contract on 1 March 2016.

Pursuant to cl 18.1(a) of the Conditions of Contract, you are hereby instructed to complete the Outstanding Works listed in the Schedule to this Certificate by 1 April 2016.

The Defects Liability Period of the Works shall commence on 1 March 2016 and will expire on 28 February 2017.

Certified by:

_____

[Name   ]
Superintending Officer

Cc [Employer]
     [QS]
     [CE]
     [ME]
     [EE]

Fig. 10.18   Certificate of substantial completion.

Schedule to the Certificate of Substantial Completion

PROPOSED CONSTRUCTION AND COMPLETION OF 4-STORY OFFICE BLOCK AT ANG MO KIO TOWN CENTRE

Outstanding Works:

(1) Plot 1: timber trellis at car park has not been completed.
(2) Gate control switch at Main Gate has not been completed.
(3) House no. has not been completed and displayed.
(4) Defects issued in Direction dated [   ] have not been rectified.
(5) ME operations manual has not been submitted.
(6) As-built drawing has not been submitted.
(7) Landscaping works have not been completed.
(8) Painting of internal doors has not been completed

The Outstanding Works shall be completed by 1 April 2016.

Fig. 10.19   Schedule to the substantial completion certificate.

10.6 Powers in Resolving Defects During Progress of Works and Defects Liability Period in PSSCOC Form of Contract

In cll 10.4, 18.3 and 18.4, the Superintending Officer has power, *during the progress of the Works and Defects Liability Period*, to require the Contractor to rectify or resolve Defects in Fig. 10.20.

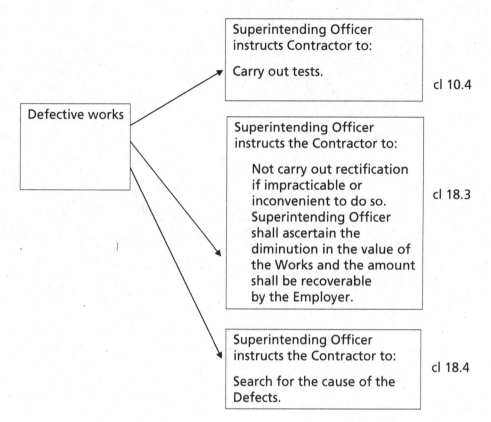

Fig. 10.20 Powers in resolving defects during progress of works and defects liability period in PSSCOC form of contract.

CHAPTER 11

Final Account

In this chapter, we will discuss the following:

11.1 What is a Final Account?

The object of a Final Account for a construction project is to compute the final Contract Sum payable to the Contractor. Prior to commencement of construction works, the Contractor would have submitted a Tender Sum for which the Owner would have accepted. The sum accepted by the Owner is known as the Contract Sum.

During the progress of the Works, there would be changes to the Works, some of which are as follows:

(a) The Architect may add more work to the Contract.
(b) The Architect may omit works from the Contract.
(c) The Architect may modify existing Contract Works.

Further, there may also be deduction from the Contract Sum as follows:

(i) Due to Defects not remedied by the Contractor, the Owner may have a right to deduct payment due to the Contractor.
(ii) Due to Delay, the Owner may deduct Liquidated Damages from the payment due to the Contractor.

Hence, the Final Account is a process to compute the Final Contract Sum payable to the Contractor after taking into account all the additions to and deductions from the Contract Sum and changes in the Works.

Naturally, the completion of the Final Account may only take place when the whole of the Works are completed and Defects remedied and resolved. This is because only then could valuation be made for all the Variation works, deduction for Defects and other adjustments made.

11.2 Final Account in the SIA Form of Contract

The time-line for submission of Final Account is in Fig. 11.1.

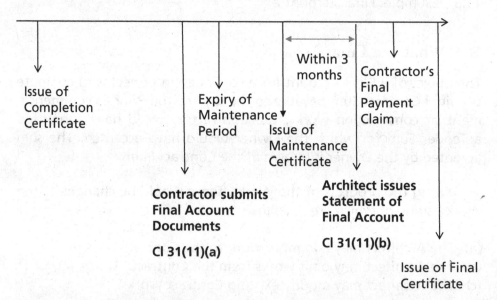

Fig. 11.1 Time-line for submission and issue of final account based on SIA form of contract.

A brief explanation in respect of the time-line is in Fig. 11.2.

S/No.	Event	Clause	Brief Explanation
1.	Contractor submits Final Account Documents	31(11)(a)	Before the end of the Maintenance Period, the Contractor submits his Final Account Documents that show the final amount which he thinks he is entitled to arising from: (a) Original Contract Works (b) Variation Works (c) NSC original and Variation Works (d) Etc...
2.	Architect issues Statement of Final Account	31(11)(b)	Within 3 months of issue of Maintenance Certificate, the Architect provides the Contractor with the Statement of Final Account, which is the Architect's final measurement and valuation of all Works.

Fig. 11.2   Brief explanation on issue of final account under SIA form of contract.

11.3 Final Account in the PSSCOC Form of Contract

The Final Account process in the PSSCOC Form of Contract in Fig. 11.3 is split into two initial time-lines (a) and (b), as indicated below. If the Contractor submits the Final Payment Claim, time-line (a) applies. If he fails to submit the Final Payment Claim, time-line (b) applies.

(a) <u>Contractor submits Final Payment Claim</u>

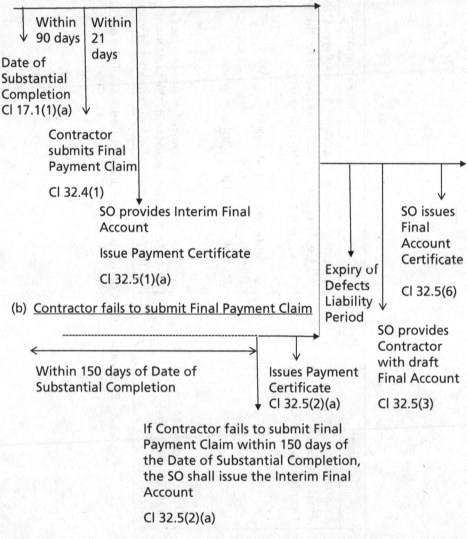

Fig. 11.3   Time-line for issue of final account in PSSCOC form of contract.

A brief explanation in respect of the time-line is in Fig. 11.4.

Item	Event	Clause	Brief Explanation
1.	In Fig. 11.3, (a) Contractor submits Final Payment Claim	32.4(1)	This part of the time-line applies when Contractor submits the Final Payment Claim. The Final Payment Claim will consist of the Contractor's Final Account Documents that show the final amount which he thinks he is entitled arising from: (a) Original contract works (b) Variation works (c) NSC original and variation works (d) Etc...
2.	Issue of Interim Final Account and Payment Certificate	32.5(1)(a)	The SO issues the Interim Final Account (separate from the Final Account in cl 32.5(6)). Many of the variations and adjustment items may be resolved in the Interim Final Account and certified in a Payment Certificate. For the Final Account, see item 4.
3.	In Fig. 11.3, (b) the Contractor fails to submit the Final Payment Claim	32.5(2)(a)	This part of the time-line applies when the Contractor fails to submit the Final Payment Claim. Within 150 days of the date of Substantial Completion, the SO issues the Interim Final Account and within 30 days, a Payment Certificate.

Fig. 11.4   Brief explanation on submission and issue of final account under the PSSCOC form.

4.	Within 30 days of the expiry of the Defects Liability Period, SO provides the Contractor with the draft Final Account	32.5(3)	Despite the adjustments in the Interim Final account and payment in a Payment Certificate, there may be other adjustments made in a draft Final Account.
5.	Exchanges between SO and Contractor in respect of the draft Final Account	32.5(4), (5)	The Contractor has an opportunity to object to the amounts in the draft Final Account, failing which he is deemed to have accepted the amounts in the draft Final account.
6.	SO issues Final Account Certificate.	32.5(6), (7)	The Final Account shows the final adjusted Contract Sum payable to the Contractor after taking into account all additions and omissions, including payments to NSC and DSC.

Fig. 11.4   (*Continued*)

11.4 Example: Final Account 1

There are many different ways of reflecting a Final Account of a construction contract. Two different formats are shown in Example: Final Account 1 and Example: Final Account 2. In Fig. 11.5, the following documents are provided for illustration:

(a) Summary of Tender
(b) Covering Letter from the Superintending Officer to the Contractor in respect of the Final Account

Summary of Tender for the Proposed Development and Construction of a Main Entry and Security Checkpoint

Item	Description	Amount ($)
1.	Preliminaries for the whole project	25,000.00
2.	Construction and completion of check-point building	155,000.00
3.	Construction and completion of vehicle barrier	30,000.00
4.	Provision and installation of air-conditioning and other mechanical works	50,000.00
5.	Provision and installation of lighting and electrical works	40,000.00
6.	Allow a provisional sum of $5,000.00 for diversion of services	5,000.00
7.	Allow a provisional sum of $5,000 for service connections	5,000.00
8.	Allow a contingency sum of $40,000 for unforeseen works	40,000.00
	Contract Sum	**$350,000.00**

Fig. 11.5(a) Summary of tender.

[Letterhead]

[Letter reference]
[Date]

XYZ Construction Pte. Ltd.   **By Email & Registered Mail**
Blk 12 #01-123 Toa Payoh
Singapore 56778
Attention: [_____]

Dear Sir,

**STATEMENT OF FINAL ACCOUNT ABC DEVELOPMENT PROJECT
PROPOSED DEVELOPMENT AND CONSTRUCTION OF A MAIN ENTRY
AND SECURITY CHECKPOINT**

CONTRACT AR 10.3

Pursuant to cl 32.5(3), Conditions of Contract, we forward herewith the Final Account in duplicate for the above-mentioned contract.

Please confirm your acceptance of the Final Account as soon as possible by having your authorized director sign and return the original copy of the Statement of Final Account. The duplicate copy is for your retention.

Kindly issue a tax invoice for the amount of $86,563.00 (including GST). Your invoice should be addressed to:

ABC Project
123 Address Road
S (123456)

Yours faithfully,

_____
Superintending Officer
Name:

cc.   [Owner]
      [QS]
      [CE]
      [ME]
      [EE]

Fig. 11.5(b)   Covering letter from the superintending officer to the contractor in respect of the final account.

(c) Statement of Final Account
(d) Final Payment due to Contractor
(e) Supporting detail to Statement of Final Account

In Fig. 11.5(a), the Summary of Tender shows the breakdown of the Contract Sum for the Proposed Development and Construction of a Main Entry and Security Checkpoint. The Contract Sum was $350,000.00. The object of a Final Account is to adjust the Contract Sum so as to arrive at the Final Contract Sum payable to the Contractor. The adjustment may be due to a variety of reasons, e.g. additional works ordered during the progress of the Works, and omission of Works from the original contract.

Fig. 11.5(b) is an illustration of a covering letter from the Superintending Officer to the Contractor, where the SO forwards the Statement of Final Account for the Contractor's acceptance.[1]

Fig. 11.5(c) shows the necessary adjustment to the Contract Sum. After the necessary adjustment to the Contract Sum, the Final Contract Sum payable to the Contractor was $330,900.00.

In Fig. 11.5(c), the Contract Sum according to the Summary of Tender was $350,000.00. The sum of $19,100 was deducted from $350,000.00 to arrive at the Final Contract Sum of $330,900.00 payable to the Contractor.

The sum of $19,100.00 to be deducted was arrived at by the omission of the Provisional and Contingency Sums of $50,000.00 in the Summary of Tender (items 6, 7 and 8) (Fig. 11.5(a)) and the addition of $30,900.00 additional Works summarized in Fig. 11.5(e).

Fig. 11.5(d) shows the Final Payment payable to the Contractor after taking into account the Final Contract Sum.

11.5 Example: Final Account 2

In Example: Final Account 2, the Nominated Sub-Contract final accounts are included in the Final Account of the Main Contractor for

---

[1]PSSCOC Form of Contract, 2014 Edition, cl 32.5(3).

[Ref:_____]
[Date   ]

STATEMENT OF FINAL ACCOUNT

PROPOSED DEVELOPMENT AND CONSTRUCTION OF A MAIN ENTRY AND SECURITY CHECKPOINT

CONTRACT AR 10.3

	$
Original Contract Sum	350,000.00

Adjustments:

Adjustment summary (See supporting details to Statement of Final Account attached)			
Ref	Omit $	Ref	Add $
Provisional and Contingency Sums	(50,000.00)	Changes due to Employer's Requirements	30,900.00
Total	(50,000.00)	Total	30,900.00

	$
	(19,100.00)
Final Contract Sum	**330,900.00**

Fig. 11.5(c)   Statement of final account.

[Ref :   ]

[Date :   ]

FINAL PAYMENT DUE TO CONTRACTOR

CONTRACT NO : AR 10.3
PROPOSED DEVELOPMENT AND CONSTRUCTION OF A MAIN ENTRY AND SECURITY CHECKPOINT

NO.	VARIATION ORDERS			
	ADDITION	NO.	OMISSION
1	$30,900.00		($50,000.00)
–		–	
–		–	
–		–	
–		–	
–		–	
TOTAL	$30,900.00	–	($50,000.00)

* Excluding GST

Contract Sum :	$350,000.00
Omission: Provisional Sums & Contingency Sums	($50,000.00)
Addition: VOs	$30,900.00
Final Contract Sum	$330,900.00
Less : Previous Payment	($250,000.00)
	$80,900.00
Less : Liquidated Damages	NIL
* Payment Due	$80,900.00
GST	$5,663.00

Fig. 11.5(d)   Final payment due to contractor statement of final account.

PROPOSED DEVELOPMENT AND CONSTRUCTION OF A MAIN ENTRY AND SECURITY CHECKPOINT (CONTRACT NO : AR 10.3)

SUPPORTING DETAILS TO STATEMENT OF FINAL ACCOUNT

Ref:	Description	Omit $	Add $	Net $
	Changes due to Employer's Requirements			
CER/1	Add new drainage channels along driveway	(0.00)	3,600.00	3,600.00
CER/2	Re-locate 2 Nos. of lamp posts	(0.00)	6,900.00	6,900.00
CER/3	Change stainless steel gates to hollow wrought iron gates	(22,000.00)	19,000.00	(3,000.00)
CER/4	Construct 1 No. of sump	(0.00)	2,000.00	2,000.00
CER/5	Add cable pipe to protect existing cable	(0.00)	500.00	500.00
CER/6	Add bitumen driveway	(0.00)	4,500.00	4,500.00
CER/7	Change culvert according to new design ref: Culvert/101	(1,500.00)	6,000.00	4,500.00

Fig. 11.5(e)   Supporting details to statement of final account.

Ref:	Description	Omit $	Add $	Net $
CER/8	Add scupper drains	(0.00)	500.00	500.00
CER/9	Add new half glaze door	(0.00)	1,500.00	1,500.00
CER/10	Add new roof gutters and rainwater down pipes	(0.00)	4,400.00	4,400.00
CER/11	Add concrete hard-standing	(0.00)	3,000.00	3,000.00
CER/12	Add 1 air-conditioning window unit	(0.00)	5,500.00	5,500.00
CER/13	Omit signage	(6,000.00)	0.00	(6,000.00)
CER/14	Re-alignment of existing driveway	(0.00)	3,000.00	3,000.00
	Total	(29,500.00)	60,400.00	30,900.00

Fig. 11.5(e)   (*Continued*)

the project Proposed Two (2) Detached Dwelling Houses on Hillside Collection at Fir Road.

In Fig. 11.6, the following documents are illustrated:

(a) Summary of Tender
(b) Statement of Final Account in respect of the Main Contract
(c) Contractor's agreement to the Final Account
(d) Adjustment for the PC and Provisional Sums
(e) Final Payment due to Contractor
(f) Supporting details to Statement of Final Account

Item	Description	Amount ($)
	Summary of Tender for the Proposed Two (2) Detached Dwelling Houses on Hillside Collection at Fir Road	

Item	Description	Amount ($)
1.	Preliminaries for the whole project	500,000.00
2.	Construction and completion of concrete footing and substructure works for detached houses A and B	258,350.00
3.	Construction and completion of detached house A	4,500,000.00
4.	Construction and completion of detached house B	3,500,000.00
5	Construction and completion of landscaping and external works	500,000.00
6	Allow a PC Sum of $225,000.00 for the supply and installation of cupboards and kitchen cabinets	225,000.00
6a.	Profit: 1%	2,250.00
6b.	Attendance	1,000.00
	Total carried forward	9,486,600.00

Fig. 11.6(a)   Summary of tender.

Item	Description	Amount ($)
	Total brought forward	9,486,600.00
7.	Allow a PC Sum of $160,000.00 for the supply and installation of kitchen appliances	160,000.00
7a	Profit: 1%	1,600.00
7b.	Attendance	1,000.00
8.	Allow a PC Sum of $90,000 for the supply and installation of lighting and electrical works	90,000.00
8a.	Profit: 2%	1,800.00
8b.	Attendance	2,000.00
9.	Allow a PC Sum of $250,000.00 for the supply and installation of air-conditioning and mechanical works	250,000.00
9a.	Profit: 2%	5,000.00
9b.	Attendance	2,000.00
10.	Allow a Provisional Sum of $500,000 for contingency works.	500,000.00
	Contract Sum	**$10,500,000.00**

Fig. 11.6(a)   (*Continued*)

Fig. 11.6(a) provides the Summary of Tender of the Main Contract. The values of the various PC and Provisional Sums were estimates provided by the Employer's QS. These values must be adjusted subsequently in the Final Account based on value of work done by the respective Sub-contractors and Main Contractor (see Fig. 11.6(b)).

[Date]

STATEMENT OF FINAL ACCOUNT

PROPOSED DETACHED DWELLING HOUSES ON HILLSIDE COLLECTION AT FIR ROAD

MAIN CONTRACT

S/No.	Description	S$
1.0	Awarded Contract Sum	10,500,000.00
	Less:	
2.0	Adjustment for PC & Provisional Sum (Annex 1)	(1,225,000.00)
3.0	Profit & Attendance (Annex 2)	(16,650.00)
	Sub-total	9,258,350.00
	Add:	
4.0	NSC's value of work done (Annex 1)	804,080.00
5.0	Profit & Attendance (Annex 2)	17,170.80
6.0	Variation to Main Contractor's work (Annex 3)	389,550.00
	Adjusted Final Contract Sum	**10,469,150.80**

Fig. 11.6(b)   Statement of final account in respect of the main contract.

In Fig. 11.6(b), adjustments are made to the awarded Contract Sum to arrive at the Adjusted Final Contract Sum of the Main Contractor. Firstly, the PC & Provisional Sums (S/No. 2.0) and Profits & Attendance (S/No. 3.0) are omitted from the awarded Contract Sum. Such values, being estimates fixed by the Employer's QS for NSC and other works, do not represent the value of work done. Secondly, NSC's actual value of work done (S/No. 4.0) and corresponding Profits & Attendance (S/No. 5.0) are added. Thirdly, the Main Contractor's value of variation works (S/No. 6.0) is added.

The Adjusted Final Contract Sum of $10,469,150.80 is the value of work done by the Main Contractor and NSCs. This sum represents the total sum payable to the Main Contractor for the *whole* of the Works, including Profit and Attendance to the Main Contractor.

During the progress of the Works, there would have been regular progress payments to the Main Contractor. Such payments should be deducted from the Adjusted Final Contract Sum before paying the balance to the Main Contractor with instructions to the Main Contractor for payment to each NSC.

If agreeable to the Final Account, the Contractor is usually required to sign his agreement for record in Fig. 11.6(c). This is a good practice as it provides certainty and clear evidence of agreement between the parties on the Final Account and leaves little room for dispute at a later date.

Fig. 11.6(d) shows the adjustment to PC & Provisional Sums. The PC and Provisional Sums provided in the Contract Sum are omitted. The awarded Sub-Contract Sums and NSC variations are added.

In Fig. 11.6(e), the Profits & Attendance of $16,650.00 in the Contract Sum are omitted. The Profits and Attendance of $17,170.80 based on the actual value of work done by the Nominated Sub-Contractors are added.

To: The Employer

We hereby agree to the Statement of Final Account dated [   ] in respect of the following project:

CONTRACT FOR THE PROPOSED DETACHED DWELLING HOUSES ON HILLSIDE COLLECTION AT FIR ROAD

Contract/Project No:[   ]

We accept the adjusted Final Contract Sum of **$10,469,150.80** in full and final settlement of the Contract. Upon payment of the outstanding sum, we will have no further claim in respect of the Contract.

We have confirmed with the Nominated Sub-Contractors and Nominated Suppliers in respect of the adjusted final contract sums for the respective sub-contracts. The Nominated Sub-Contractors and Nominated Suppliers do not have any further claim. We undertake to pay the outstanding amounts to the Nominated Sub-Contractors and Nominated Suppliers upon receiving payments from the Employer.

Notwithstanding our agreement to the adjusted Final Contract Sum, we understand that the Employer reserves his right to set-off or recover any Defects or Liquidated Damages or for any other loss and damages provided in the Contract or at law.

_____

Name: [   ]
Designation: [   ]
For and on behalf of:
               [   ]

Company stamp: [   ]

Date: [   ]

Fig. 11.6(c)   Contractor's agreement to the final account.

Annex 1

Adjustment for PC & Provisional Sums

S/No.	Description	Omission $	Addition $
A	Nominated Sub-Contract		
1.	PC Sum for the supply and installation of cupboards and kitchen cabinets in Item 6, Summary of Tender.	(225,000.00)	
	Awarded Sub-contract sum (see Fig. 11.6(g))		440,000.00
	Variations		4,040.00
2.	PC Sum for the supply and installation of kitchen appliances in Item 7, Summary of Tender	(160,000.00)	
	Awarded Sub-contract Sum (see Fig. 11.6(h))		47,040.00
	Variations		NIL
3.	PC Sum for the supply and installation of lighting and electrical works in Item 8, Summary of Tender	(90,000.00)	
	Awarded Sub-contract Sum (see Fig. 11.6(i))		86,000.00
	Variations		(5,000.00)
4.	PC Sum for the supply and installation of air-conditioning and mechanical works in Item 9, Summary of Tender	(250,000.00)	
	Awarded Sub-contract Sum (see Fig. 11.6(j))		210,000.00
	Variations		22,000.00

Fig. 11.6(d)   Adjustment to PC & provisional sums.

S/No.	Description	Omission $	Addition $
5.	Provisional Sum for contingent works in Item 10, Summary of Tender	(500,000.00)	
	Adjustment to PC & Provisional Sums	1,225,000.00	804,080.00

Fig. 11.6(d)   *(Continued)*

<u>Annex 2</u>

Adjustment for Profit & Attendance

S/No.	Description	Final Sub-Contract sum $	Omission $	Addition $
A	<u>Nominated Sub-Contract</u>			
1.	Nominated Sub-Contract for the supply and installation of cupboards and kitchen cabinets PC Sum: $225,000.00	444,040.00		
	Profit (1%) Attendance		(2,250.00) (1,000.00)	4,440.40 1,000.00
2.	Nominated Sub-Contract for the supply and installation of kitchen appliances PC Sum: $160,000.00	47,040.00		
	Profit (1%) Attendance		(1,600.00) (1,000.00)	470.40 1,000.00

Fig. 11.6(e)   Adjustment for profit & attendance.

S/No.	Description	Final Sub-Contract sum $	Omission $	Addition $
3.	Nominated Sub-Contract for the supply and installation of lighting and electrical Works PC Sum: $90,000.00	81,000.00		
	Profit (2%)		(1,800.00)	1,620.00
	Attendance		(2,000.00)	2,000.00
4.	Nominated Sub-Contract for the supply and installation of air-conditioning and mechanical Works PC Sum: $250,000.00	232,000.00		
	Profit (2%)		(5,000.00)	4,640.00
	Attendance		(2,000.00)	2,000.00
			16,650.00	17,170.80

Fig. 11.6(e)   (*Continued*)

Annex 3

Variation to Main Contractor's work

S/No.	AI No.	Description	$
1.	1	Enlarge the concrete footing	5,550.00
2.	-	Adjust the piling works	(30,000.00)
3.	2	Change to pile-cap size	7,500.00
4.	3	Add drainage	10,500.00
5.	4	Change door height	11,000.00
6.	5	Change brick wall to demountable partition	35,000.00
7.	6	Change cement & sand screed to ceramic floor tile	45,000.00
8.	7	Change painting to wall tile in toilets	60,000.00
9.	8	Change wash basin in toilet to vanity top	70,000.00
10.	9	Change chain-link fencing to ornamental fencing as described I AI.	80,000.00
11.	10	Change window grilles to ornamental window grilles as in AI.	75,000.00
12.	11	Change entrance gate to remote control electronic gate as in AI.	20,000.00
			389,550.00

Fig. 11.6(f)   Variation to main contractor's work.

In Fig. 11.6(f), with reference to each Architect's Instruction (AI), the figure tabulates the value of Variation works of the Main Contractor. Further details of measurements (take-off quantities) and rates should be provided in separate sheets.

Fig. 11.6(g) to Fig. 11.6(j) show the Statements of Final Accounts for the Nominated Sub-Contractors.

Arising from the Final Account in Fig. 11.6(b), the Final Certificate is certified to the Main Contractor in Fig. 11.7. Corresponding instructions should be issued to the Main Contractor for payment to the Nominated Sub-Contractors arising from the Final Certificate.

Statement of Final Account for Nominated Sub-Contract for the supply and installation of cupboards and kitchen cabinets.

Awarded Sub-Contract Sum $440,000.00

Variations:

Description	Omission	Addition
AI 12 Additional cupboard model 12A	NIL	4,040.00
		4,040.00

Adjusted Final Sub-Contract Sum: $444,040.00

To: [Main Contractor]
    [Employer]

We hereby agree to the Statement of Final Account in respect of the following project:

Project: **Nominated Sub-Contract for the supply and installation of cupboards and kitchen cabinets**

Contract/Project No: [   ]

We accept the adjusted Final Sub-Contract Sum of $444,040.00 in full and final settlement of the Sub-Contract. Upon payment of the outstanding sum, we will have no further claim in respect of the Sub-Contract.

Name: [   ]
Designation: [   ]
For and on behalf of :
[   ]
Company stamp: [   ]
Date: [   ]

Fig. 11.6(g) Nominated Sub-Contractor's Final Account for the supply and installation of cupboards and kitchen cabinets.

Statement of Final Account for Nominated Sub-Contract for the supply and installation of kitchen appliances.

Awarded Sub-Contract Sum $47,040.00

Variations:

Description	Omission	Addition
NIL	NIL	NIL

Final Sub-Contract Sum: $47,040.00

To: [Main Contractor]
    [Employer]

We hereby agree to the Statement of Final Account in respect of the following project:

Project: <u>Nominated Sub-Contract for the supply and installation of kitchen appliances</u>

Contract/Project No: [   ]

We accept the adjusted Final Sub-Contract Sum of $47,040.00 in full and final settlement of the Sub-Contract. Upon payment of the outstanding sum, we will have no further claim in respect of the Sub-Contract.

_____

Name: [   ]
Designation: [   ]
For and on behalf of:
[   ]
Company stamp: [   ]

Date: [   ]

Fig. 11.6(h)   Nominated Sub-Contractor's Final Account for the supply and installation of kitchen appliances.

Statement of Final Account for Nominated Sub-Contract for the supply and installation of lighting and electrical works.

Awarded Sub-Contract Sum $86,000.00

Variations:

Description	Omission	Addition
AI 13 Omit one (1) circuit breaker	(5,000.00)	NIL
	(5,000.00)	NIL

Final Sub-Contract Sum: $81,000.00

To: [Main Contractor]
    [Employer]

We hereby agree to the Statement of Final Account in respect of the following project:

Project: <u>Nominated Sub-Contract for the supply and installation of lighting and electrical works</u>

Contract/Project No: [   ]

We accept the adjusted Final Sub-Contract Sum of $81,000.00 in full and final settlement of the Sub-Contract. Upon payment of the outstanding sum, we will have no further claim in respect of the Sub-Contract.

Name: [   ]
Designation: [   ]
For and on behalf of:
[   ]
Company stamp: [   ]

Date: [   ]

Fig. 11.6(i)   Nominated sub-contractor's final account for the supply and installation of lighting and electrical works.

Statement of Final Account for Nominated Sub-Contract for the supply and installation of air-conditioning and mechanical works.

Awarded Sub-Contract Sum $210,000.00

Variations:

Description	Omission	Addition
AI 14 additional six (6) Daikin air-con window units	NIL	22,000.00
		22,000.00

Final Sub-Contract Sum: $232,000.00

To: [Main Contractor]
     [Employer]

We hereby agree to the Statement of Final Account in respect of the following project:

Project: <u>Nominated Sub-Contract for the supply and installation of air-conditioning and mechanical works</u>

Contract/Project No: [   ]

We accept the adjusted Final Sub-Contract Sum of $232,000.00 in full and final settlement of the Sub-Contract. Upon payment of the outstanding sum, we will have no further claim in respect of the Sub-Contract.

Name: [   ]
Designation: [   ]
For and on behalf of:
[   ]
Company stamp: [   ]

Date: [   ]

Fig. 11.6(j)   Nominated sub-contractor's final account for the supply and installation of air-conditioning and mechanical works.

FINAL CERTIFICATE NO. [   ]
[Date]

**Project: PROPOSED DETACHED DWELLING HOUSES ON HILLSIDE
COLLECTION AT FIR ROAD**

Contractor: Best Contractor Pte. Ltd.
Address: 888 Kampong Bahru Road, #01-01 Infrastructure Industrial Estate,
Singapore 654321

Pursuant to Clause 31 (12) of the Conditions of Contract, I hereby certify
payment to the Contractor as follows:

Value of work done by Contractor	$9,665,070.80
Value of materials on site	$ -
Value of work done by Nominated Sub-Contractors	$ 804,080.00
Total	$10,469,150.80
Less: Sums previously certified	($8,500,000.00)
Amount due (excluding GST)	$1,969,150.80

Certified by:

_____

Name:
Architect

Distribution to:

Fig. 11.7   Final Certificate for payment of balance in final account.

Index

Printed in the United States
By Bookmasters